Spirits and Clocks

Spirits and Clocks

Machine and Organism in Descartes

Dennis Des Chene

Cornell University Press

Ithaca and London

First published 2001 by Cornell University Press

Printed in the United States of America

Library of Congress Cataloging-in-Publication Data

Des Chene, Dennis.
    Spirits and clocks : machine and organism in Descartes / Dennis Des Chene.
        p.   cm.
    Includes bibliographical references (p.   ) and index.
    ISBN 0-8014-3764-4 (cloth : acid-free paper)
        1. Descartes, René, 1596-1650—Knowledge—Physiology.
    2. Physiology—Philosophy.   I. Title.
    QP21 .D378   2000
    571′.01—dc21

                                                                    00-009872

Cornell University Press strives to use environmentally responsible suppliers and materials to the fullest extent possible in the publishing of its books. Such materials include vegetable-based, low-VOC inks and acid-free papers that are recycled, totally chlorine-free, or partly composed of non-wood fibers. Books that bear the logo of the FSC (Forest Stewardship Council) use paper taken from forests that have been inspected and certified as meeting the highest standards for environmental and social responsibility. For further information, visit our website at www.cornellpress.cornell.edu.

        1   3   5   7   9   Cloth printing   10   8   6   4   2

To my mother and father
Dorice Marie Des Chene
Raymond Joseph Des Chene

CONTENTS

Contents

FIGURES

PREFACE

Like its companion, *Life's Form,* this work is a historical study of life, organism, and unity in early modern natural philosophy. It began with vague thoughts that Descartes takes the intelligibility of machines for granted, and treats them as individuals. Yet he has no definition, nor any account of unity for them. Those thoughts mingled with others, inspired by Canguilhem and more remotely by Kant, on life and unity. I was drawn from topics properly psychological and centered on the human case to fundamental questions concerning soul in general. In the Aristotelian classification of the sciences, the science of the soul has charge of all the phenomena of life. It embraces not only the lowliest plants and animals, but also purely spiritual beings, including God. In Descartes' work, on the other hand, the relation of the soul (which exists only in humans) and of God to vital operations seems to have been severed—a separation with grave effects for body and soul alike.

I began to think that the changes that befell the concept of life and concepts allied with it were as momentous as those undergone by *mind* or *corporeal substance.* Not a new thought, certainly. Yet the mass of papers and books on Cartesian physics and psychology far outweighs the small number on his physiology. There is of course no dearth of work on the animal-machine. But most of it takes up Descartes' claim that machines cannot think, or interprets the theory of the senses laid out in the *Traité de l'Homme.* His animal-machine becomes a stand-in for the automata of our time. But can it be of no importance that the Cartesian soul, unlike the Aristotelian, has no role in the vital operations of the body to which it is joined? What was the import of the act of calling animals machines? If in *Le Monde* Descartes was rewriting Aristotle's

Physics, was he also in *L'Homme* rewriting *De Anima?* Those are the questions that interested me.

Descartes' physiology includes much that is unworkable, even incredible. For more than a few of his successors and for some historians it became an embarrassment, a misbegotten adventure to be passed over hastily. Yet his rejection of the vegetative and sensitive souls invented by his predecessors was no less momentous for the sciences of life than the rejection of forms, powers, and ends was for physics. It was of a piece with his program in physics, executed with the same motives and resting on the same principles.

The *De Anima* Descartes knew is not quite Aristotle's work as we understand it. It is that work as it was presented in Descartes' time, with the commentaries and *quæstiones* added to it by early modern Aristotelians—in particular the Jesuits from whom Descartes learned his Aristotle, and with whom he struggled in later life. In *Life's Form* I studied some of the principal questions on the soul in general, drawing mainly on Jesuit authors—the Coimbrans, Suárez, Arriaga. Here I treat similar questions in Descartes. I make no claim to comprehensiveness in either work. A wealth of topics remains to be explored in Aristotelian authors—including the Dominican Thomists and Scotists who were contemporaries of, and sometimes opponents of, the Jesuits I have looked at. Separated souls, the plurality of forms espoused by Scotists, or the use of medical texts in late Scholasticism: it is not difficult to find areas as yet hardly cultivated. As for Descartes, I have not dwelt on his medical ideas, or his use of medical authors. Much more could be done with the notion of machine in the period, and of course Descartes' work on generation should be placed in the context of contemporaries like Gassendi and Regius.

I thus regard *Life's Form* and the present work as prolegomena to a fuller and deeper history to come. Among those who helped me with this initial segment are my former colleagues George Wilson, Jerry Schneewind, Susan Hahn, and Karen Neander; Adam Goldstein, my occasional research assistant and sounding-board; Roger Haydon of Cornell Press, ever receptive to nascent projects; Peter Galison and John Murdoch, at whose invitation I spent an invigorating semester at the Harvard History of Science Department just as I was beginning the manuscript from which *Life's Form* and *Spirits and Clocks* have issued; Roger Ariew, Dan Garber, Allan Gabbey, Stephen Menn, Alison Simmons, Helen Hattab and Emily Michael, fellow Descartes scholars whose publications and conversations have encouraged and guided me; Stephen Gaukroger, who invited me to a conference on Descartes' natural philosophy at Sydney in 1996 at which I presented a paper ancestral to *Spirits and Clocks;* Joël Biard, the organizer of the conference "Descartes et le Moyen Âge" at the Sorbonne in 1996; the National Endowment for the Humanities (for a Fellowship in 1996) and the American Council of Learned Societies (for a travel grant); and finally my wife Mary, who every day reminds me both that philosophy matters and that it is not all that matters.

Spirits and Clocks

INTRODUCTION

The world of Descartes' physics is austere enough to gladden the heart of the most fervent reductionist. Cartesian matter is space itself, and all that pertains to matter is no more than an elaboration of figure, size, and motion. From that bare inventory is produced everything that nature presents to our senses: the stars and planets; the rivers, seas, and mountains of our planet; its metals, stones, and oils, its plants, its animals, the human body. All forms, all powers, all qualities other than the modes of matter are evicted from the world of Descartes. Its plants and animals have no souls: nothing sets them apart from the inanimate world around them. Only in humans is there a distinct principle, a substance whose modes are not found among, or derived from, the modes of matter.

The elimination of the souls of animals and plants is of a piece with the elimination of forms and qualities generally from nature. Since other living things, with their great variety of visible forms and actions, and their numerous similarities to humans, offer the greatest resistance to the Cartesian program, physiology,[1] even more urgently than physics, had to go under the

---

[1] 'Physiology' (*physiologia*) was originally used to denote all of natural philosophy, and is so used in its infrequent occurrences in Aristotelian textbooks. In the sixteenth century it began to be used to denote the "natural part" of medicine—the study of the nature, powers, and functions of human beings. In this sense physiology included anatomy. 'Physiology' is used in the modern sense in the middle of the eighteenth century, although the older use as a synonym of 'natural philosophy' continued in Germany to the end of that century. See Rothschuh 1968:13–14, Duchesneau 1982:xiii–xiv. The Aristotelian science of the soul included both physiology and psychology in the modern sense; though it is slightly anachronistic, I will sometimes use 'physiology' to denote the part of that science that studied the operations of the vegetative soul, and also the corresponding part of Descartes' work.

knife if the program was to succeed. In certain respects this was the more radical operation. Those who, like Descartes' Aristotelian predecessors, defended forms and qualities might well concede that in physics they were superfluous. But that living things should lack souls—that they should be nothing other than machines—was, for many of Descartes' contemporaries, an insuperable obstacle to the acceptance of his natural philosophy.

Some aspects of Descartes' work on living things have been studied: the theory of the senses and the passions, the strategy of simulation proposed in *L'Homme,* the criteria by which to distinguish bodies with souls from mere automata.[2] Descartes' physiology, by contrast, has often been given only cursory treatment or dismissed.[3] Yet he devoted about one-fourth of *L'Homme* to circulation, locomotion, respiration, and digestion;[4] a significant portion of the fifth part of the *Discours* reports on the work in *L'Homme;* the *Description du corps humain* is entirely devoted to physiology and embryology; and various notes and portions of his letters take up particular questions, notably the role of the heart in circulation. He invested more time trying to unravel the mysteries of generation than in explaining the behavior of light, and (if one judges by his advice to Queen Elisabeth) much more than in pondering the mysteries of metaphysics. Add to this a concern, amounting to obsession according to some contemporaries, with health and the prolongation of life, and it is clear that the neglect of Descartes' physiology does not accord well with the portion of his working life that was given over to it. It reflects, rather, philosophers' current interest in the cognitive faculties and in questions of method.

Yet the revolution that in *L'Homme* was proposed for the science of life was if anything more radical than the revolution in physics proposed in *Le Monde.* Descartes proposes to eliminate the living as a natural kind. The science of life is henceforward to be, not the science of a special part of nature consisting in those things that live, and that therefore have souls, but an extension of

[2] Among recent works, see Perler 1996 on sensory representation, Kambouchner 1995 on the passions, and Giglioni 1995 on automata.

[3] Recent exceptions to the rule include Bitbol-Hespériès 1990 and Grene 1993; Pichot 1993 is a guide to the texts with an insightful commentary; the new edition of *Le Monde* and *L'Homme* edited by Bitbol-Hespériès and Verder is indispensable.

[4] In the table of contents supplied by Clerselier for his edition of *L'Homme* (1664), two of the five parts and 26 of 106 articles are devoted to physiology. The *Description,* written in 1648, was published by Clerselier with *L'Homme.* The *Primæ cogitationes de generatione animalium,* which are a series of fragments from various periods of Descartes' career, were published only in 1701, and the *Anatomica,* which are excerpts by Leibniz from Descartes' manuscripts, in 1859–1860. It should be noted that despite the posthumous publication of all but the *Discours,* Descartes' views were circulating already in the 1640s by way of Regius and others.

physics.[5] Descartes' *Principia,* had it been completed according to plan, would have moved without a break from explaining the behavior of magnets to explaining the behavior of plants. Aristotelian natural philosophy interpolated a series of forms, of increasing perfection, between prime matter and the immortal soul. Descartes' natural philosophy has only two levels of perfection: extended substances, which are all on a par, and the soul. He does take over the characterization of animals as self-movers or *automata.* But like other verbal coincidences, this one hides a shift of concept. The principle of life is no longer a form peculiar to living things: it is the heat of the heart, the same heat which is found in fire or rotting hay, and which consists in the violent motion of small particles. Though we are inclined to believe that animals, because they resemble us in so many ways, must have souls like ours, in fact there is nothing in them that resembles the soul—no sensations, no passions, no judgments or volitions.

Conversely, the human soul has nothing to do with the vital operations that in Aristotelian physiology are referred to the vegetative part of the soul. These belong instead to the body-machine. In Aristotelianism the domain of the living stretches from the lowly plant to the perfect being; in Cartesianism there are two separate domains, joined only by way of the union of the human soul and its body. In neither do you find living things in the Aristotelian sense. The body-machine does not live, since it has no powers, but only passive qualities derived from the modes of extension; nor does the soul, since it has no part in nutrition, growth, or generation.

Descartes' program, then, is to explain all those functions of the body that occur in us without thought. Those functions include the functions of the vegetative part of the soul as well as those of the sensitive soul: nutrition, growth, and generation, on the one hand; sensation, passion, imagination, memory, and locomotion on the other. Of the vegetative functions, nutrition and growth were easily disposed of by adapting the accounts of his predecessors. But generation proved to be an obstacle Descartes never quite managed to overcome. In the absence of any organizing power, such as others thought to be present in the seed, the formation of the fetus must result from the operation of efficient causes acting independently of each other and of the final form. Thus is the vegetative part of the soul suppressed, together with those functions of the sensitive soul—what we would now call "reflex actions," for example—that do not require thought. The living world, humans aside,

[5] "No epistemological frontier separated the part of physics that occupied itself with inorganic realities from that which occupied itself with organic realities" (Duchesneau 1982: xiv). The overlap in the two uses of 'physiology' mentioned in n. 1 is indicative of the lack of distinction.

has no property, and includes no entity, that would distinguish it from the nonliving.

Part I of *Spirits and Clocks* treats first the question of the principle of life, and then the animal-machine as self-mover. Explaining the cyclical operations of the body evoked from Descartes a variety of new or adapted concepts—the system by which the pulse is produced, the fluidity of the body, the *Ur-animal* or basic body plan which is supposed to result from the first circular flow of the blood. Though there is only a hint at an explanation of the transmission of characters, Descartes does attempt to explain individual differences. In particular he offers, in Latin notes not published in his lifetime, two accounts of sexual difference. All these phenomena are to be explained without reference to ends. The banishing of ends from Cartesian physiology raises two questions: the basis upon which parts or organs in the body are to be designated, and the role of normality or health in his project.

The "statues" of *L'Homme* are said to be machines, constructed by God so as to imitate human bodies as closely as possible. But it is clear that those machines are not just *like* human bodies: they *are* human bodies. That notorious claim raises a number of questions, which occupy Part II of this work. What was the import in Descartes' time of calling something a machine? "Machinism" was a tool by which to bring living things within the scope of a mechanistic physics. It enabled the application to them of the analysis of capacities which the engineers of the period were beginning to codify in their descriptions of machines—the fountains and pipe organs that Descartes asks his reader to consider in *L'Homme,* for example. In principle, the analysis of capacities would break down the organism into simple mechanisms whose operations could be explained by the laws of nature. Ideally, it provides a bridge between the behavior of complex machines and the realm of the demonstrable in physics; in practice, it was a device for persuading readers that a bridge existed—if only in the mind of God.

The machine is, stereotypically, an artifact, something made. Christian philosophy had long regarded God as the *artifex maximus* and his creations as analogous to human art. There was, in that respect, nothing new in Descartes' comparison of animals and machines. But Aristotelian philosophers instituted a difference in productive power between divine creation and human industry. God alone, or his spiritual intermediaries, can create the souls of the more perfect things in nature, especially animals. The formation of the body and its organs belongs to the seed and the womb; but the soul, which is the form of the body, must be introduced by a higher agent. Descartes, on the other hand, would have it that animals differ only in degree of complexity from the clocks and pumps we make. Their parts are finer, more intricate, but not in any sense more perfect. Unless, as Leibniz would later argue, God's artifacts are endowed with an *infinite* complexity, there is no essential distinction between his power to make animals and ours to make machines. One marker on the boundary between nature and art is thereby erased.

Descartes' earlier works—from the notes written in 1619–1620 to *L'Homme*—dwell on the possibility of *simulation*. Machines can produce illusions; they can also dispel them. In combatting our inveterate opinion that animals are not machines, Descartes urges that we imagine machines capable of simulating all the operations we see in animals, which presupposes the existence of the things simulated. On the other hand, he invites us to consider a world in which there are no animals, but only machines that more or less closely resemble the animals of the actual world. The machine-simulation would deceive us were we to encounter it, *except* that (paradoxically enough) since the animals of the actual world *are* machines, we would not, as it turned out, be deceived at all. The argument here turns on two sorts of resemblance: the *good* resemblance between various familiar mechanisms and the organs of animals, and the *bad* resemblance between the organs and operations of animals and our own. The good resemblance leads to truth, the bad to falsehood. I will consider, in Chapter 5, how Descartes manages to distinguish good resemblance from bad, and the role of simulation in the project of his physiology.

If one takes Descartes at his word, the only material individuals in nature are either regions in space or else collections of such regions moving together in one direction. Call this "physical" unity. The body-machine evidently lacks physical unity. Yet Descartes, like everyone else, speaks of the body as if it were one thing. Even the machines of *L'Homme* are so treated, despite the absence of souls in them. What then is the principle of unity of the body? In Chapter 6 I consider the various kinds of unity a Cartesian machine, animal or human, might be said to have. In addition to physical unity, a machine has that unity which consists in the joint operation of its parts to produce, under certain conditions, a single effect or range of effects—this I call "dispositional" unity. Commonsensically, its organs also have functional unity—the eye is one thing by virtue of its power to see. Though Descartes uses 'function,' 'office,' and other allied terms, it is not clear that he has a basis for doing so other than our ingrown tendency to project purposes onto the objects of nature. Useful though that tendency might be in everyday life, Descartes excludes it from natural philosophy. For other philosophers of his time divine intentions provided a basis for ascriptions of functions: the functional unity of an organ rests on the intentional unity it has by virtue of having a purpose assigned to it by God. But consideration of divine intentions is, in all but the human case, again excluded by Descartes from natural philosophy.

One last sort of unity is substantial unity. In Aristotelian physiology the body of an animal can be regarded as one thing by virtue of its union with a substantial form—the soul. Descartes occasionally asserts that for him too, in the human case, the soul is the substantial form of the body. In that case, then, but in that case alone, the body has substantial unity. But Cartesian soul has nothing in it that would imply that it should be joined with a *human* body, rather than the body of a rat, or for that matter a piece of bread. What remains puzzling is that we should have the bodies we have, that so complicated a machine should be joined with a soul.

A few words, finally, about the ambitions of this work. I have not undertaken here a comprehensive study of the literature on Cartesian machines, and still less on machines generally in early modern philosophy. My intention has been to study what are, after all, familiar texts from a slightly different point of view, emphasizing the noncognitive functions of living things, those which are shared by all. The human case retains its centrality, if only because Descartes wrote very little on animals and almost nothing on plants. But I think it fruitful to move away from an exclusive focus on cognitive powers and on the mind-body problem, and to situate Descartes' work in the history of physiology rather than in the history of psychology alone. It was, after all, his relegation of the vital powers of the soul to the body alone that helped bring about the separation of those two disciplines.

∞

A note on references. Wherever possible, for primary sources I have given references to the smallest identified parts of a work—chapters, sections, paragraphs, parts of questions, as well as page numbers. In page references, 'r' and 'v' denote recto and verso; 'a' and 'b' denote the first and second column on a page. For parts of texts I use the following abbreviations: 'a' or 'art' for 'article', 'ad' for subsection of an article, 'c' for 'chapter', 'disp' for disputatio, 'lib' for 'book' ('liber'), 'prop' for proposition, 'q' for 'quæstio', § for numbered sections, 'tex' for 'textus'. Primary texts are identified by author and (abbreviated) title, with the editor's name if necessary; secondary texts by author and year; with reprints I use the original date of publication for primary texts; for secondary texts, that and the date of publication of the version I refer to (the original date is usually given in the bibliography). In Latin quotations I have usually changed 'i' to 'j' and 'u' to 'v' in accordance with typographical conventions now, but I have, in all but one or two instances, left punctuation as is. Abbreviations other than those in citations and the ampersand have been expanded. All translations, unless otherwise indicated, are mine.

PART ONE

TALES OF THE *BÊTE-MACHINE*

> I spent a winter in Amsterdam, where I went almost every day to the
> house of a butcher, to see him kill the animals, and had brought to my
> lodgings the parts I wanted to anatomize at greater leisure.[1]
> —Descartes to Mersenne, 13 Nov 1639

> The animal, according to certain authors, is a hydraulic machine.
> What stupid things can be said on the basis of that one supposition![2]
> —Diderot

The *Traité de l'Homme* starts abruptly: "These men will be composed, like us,
of a Soul and a Body."[3] They will be "of such a nature that when their eyes
are pushed on" by light, "they will have a sensation of it entirely resembling
the sensation we have" (AT 11:97). The body of these men will be "nothing
other than a statue or machine of earth, which God has formed expressly to
render it as similar as possible to us" (120).[4] Their world will be like ours, their

---

[1] "j'ai été un hiver à Amsterdam, que j'allais quasi tous les jours en la maison d'un boucher,
pour lui voir tuer des bêtes, & faisais apporter de là en mon logis les parties que je voulais
anatomiser plus à loisir" (AT 2:621).

[2] "L'animal, suivant quelques auteurs, est une machine hydraulique. Que de sottises on peut
dire d'après cette unique supposition!" Diderot *Élém. physiol.* ptic2, p21.

[3] AT 11:119–120. According to Clerselier, who edited the first French edition of the *Traité
de L'Homme* in 1667, the beginning of that work is designated "Chapitre XVIII." Since the preceding work, the *Traité du Monde,* ends with Chapter 15, two chapters were evidently missing
even then.

[4] Summarizing *L'Homme* in the *Discours,* Descartes writes: "je me contentai de supposer que
Dieu formât le corps d'un homme, entièrement semblable à l'un des notres tant en la figure ex-
térieure qu'en la conformation intérieure de ses organes, sans le composer d'autre matière que
celle que j'avais décrite, et sans mettre en lui, au commencement, aucune âme raisonnable, ni
aucune autre chose pour y servir d'âme végétante ou sensitive, sinon qu'il excitât en son cœur

light, their bodies, even their sensations. But we know full well, from Descartes' other works, that the screen of resemblance erected between the world of the *homme-machine* and our own is a kind of feint. Our world is not just *like* the world of *Le Monde*, it *is* that world, and the people in our world *are* the people there.

For now, therefore, I disregard the distinction between *is like* and *is,* and take the *Traité de l'Homme* to be a treatise on man, not something like man. So taken, it offers both less and more than its title suggests. Less, because there is nothing about generation, and the promised parts on the soul and the union do not, and perhaps never did, exist. We learn something of the soul in the *Meditations* and the *Objections and Replies,* and something of the union there and in the *Passions of the Soul.* But even the *Principles,* though they absorb much of *Le Monde,* and a bit of *L'Homme,* lack the parts that would have treated soul and union. Descartes never wrote, or saw fit to preserve, any work devoted exclusively to them.[5]

The *Traité de l'Homme* offers more than its title implies because the human body is the type of all animal bodies.[6] Not even the fountain-works in the brain that give us fluid purchase on the world are peculiar to humans. They serve in us also for "all those functions that can be imagined to proceed from matter, and to depend only on the disposition of the organs" (120). There is no reason to think that in animals those functions are not subserved by similar machinery. *L'Homme,* despite its name, offers a theory of animal life to rival that of *De Anima.*

On the phenomena of life Descartes differs little from the Aristotelians,

---

un de ces feux sans lumière." (*Discours* 5, AT 6:45–46; cf. *L'Homme* AT 11:120, 202). But the *Discours* goes on to defend Descartes' theory of the heart, and it is clear that *L'Homme* is, no less than the *Essais* that accompanied the *Discours,* intended to be literally true of our bodies. On the *feu sans lumière,* see §1.2 below.

[5] On the parallels between *Le Monde* and the *Principles,* see AT 11:704–706. On the composition of *Le Monde* and *L'Homme,* see AT 11:iii–v and Clerselier's preface, ibid. xix, xxiii–xxiv; Bitbol-Hespériès in *Le Monde/L'Homme* 171–172n4. No autograph of *Le Monde* or *L'Homme* now exists; according to Clerselier, the second edition of *Le Monde* (1677) was based on an autograph. Neither is mentioned in the inventory of Descartes' works taken after his death in Stockholm. Bitbol-Hespériès suggests that Descartes did not take the manuscripts with him because they "did not coincide entirely with the revised summary he had proposed for them in the *Discourse*" (Bitbol-Hespériès, ibid.). Though the summary in the *Discours* mentions descriptions of plants and animals and the rational soul (AT 6:59), the works contain none.

[6] Some years after ceasing to work on *L'Homme,* Descartes writes, "And if I were to begin again my *World,* where I supposed the body of an entirely formed animal, and contented myself with showing its functions, I would undertake to put in also the causes of its formation and birth [. . . .] I aim to know the animal in general [. . .] and not yet man in particular." (To Mersenne 20 Feb 1639; AT 2:525–526).

who had already made many of the necessary accommodations to Galenism.[7] He admits the circulation of the blood, newly discovered by Harvey, though he disagrees with Harvey on the role of the heart. The dissections and experiments on animals Descartes performed, though unlikely to have been performed by the authors of *De Anima* commentaries, were by no means unprecedented.[8] From Kepler he takes the description of the eye as a *camera obscura,* making the retina, and not the crystalline humor, the site of visual sensing. But on the whole, there is little novelty in his tabulation of the things to be explained,[9] nor in the principles by which he explains them, notably the animal spirits.[10] The strategy of *L'Homme* is not that of Galileo announcing new planets and mountains on the Moon, or of Gassendi slashing away at the errors of Aristotle. *L'Homme* promotes a revolution that is *conservative* over the phenomena, but *radical* in its interpretation of them. What matters is that the whole physical world should be brought under Descartes' new principles. The human body, and by implication those of animals, belong to that world as surely as Aristotelian human beings and animals belong to Aristotle's.

Life, whatever it may be, must be a feature of bodies that have no other modes than those of substance in general and of extended substance in par-

---

[7] If Descartes read Galen at all, he most likely read him in Latin translation. Mesnard finds it "very probable" that Descartes knew Galen only through Jacobus Silvius (Jacques Dubois): see Sylvius *Isagoge* (1560), *Opera* (1630) (Mesnard 1937:208). For an extended comparison of the physiology of *L'Homme* and the physiology of Galen, see Hall's translation of *L'Homme.*

[8] Among the excerpts made by Leibniz from Descartes' notes, one excerpt containing miscellaneous observations on animals and humans is dated 1631 (AT 11:601). From Descartes' correspondence, we know that he was studying anatomy and dissecting animals in 1629–1630 (To Mersenne 15 Apr 1630, AT 1:137; Nov or Dec 1632, AT 1:263; 20 Feb 1639, AT 2:525; 20 Nov 1639, AT 2:621). He may well have observed human dissections at the famous anatomical amphitheater at Leiden (see Cavaillé 1991:17–30, esp. 17n1).

[9] Contrasting himself with a doctor "who would have it that the valves of the heart do not exactly close," Descartes writes, "I have supposed nothing new in Anatomy, nor anything that is in any way controversial among those who write on the subject" (To Mersenne 14 Jun 1637, AT 1:378). He is circumspect about, but does not reject, some of the dodgier *experientia* of natural history, like the *marques d'envie* impressed on fetuses by their mothers, or the order of battle said to be observed by apes. On the phenomena of life in Aristotelian texts, see Des Chene 2000, Chapter 1.

[10] See Pichot 1993, c5.2:342–387. As for the heat of the heart, Plempius points out to Descartes that "your opinion is not new, but old, and indeed Aristotelian," citing *De respiratione* 20. Aristotle there says that "pulsation [. . .] is the inflation of the humor as it is heated" ("Pulsatio igitur est humoris concalescentis inflatio") (Plempius to Descartes Jan 1638, AT 1:496). In the *Description,* Descartes, citing Aristotle, holds that it was only "by chance that Aristotle managed to say something approaching the truth" (*Descrip.* 2, AT 11:245). The harshness of this judgment contrasts markedly with the almost obsequious praise in his answer to Plempius, tinged with irony though it is ("I would wish for nothing more than to be able to follow, while not slipping away from truth, his track in all things") (Descartes to Plempius Jan 1638, AT 1:522).

ticular. Bodies endure, they are extended in height, width, and depth, they move. They cannot move themselves, or rather they cannot change their own motions; in particular no body can bring about its own motion. How, then, to characterize them, or to explain the commonsense distinction between the living and the nonliving? How explain even the appearance of self-motion?[11] Aristotelian authors tend to pick out certain marks peculiar to the operations of living things—Arriaga's *intus-sumption,* the list of marks by which Toletus distinguishes vital from nonvital generation.[12] But those marks are not decisive. Suárez, for example, agrees that the operations of living things could be simulated, not by machines, but by God. What makes an operation vital is that it is referred to, and "denominated from," as Suárez puts it, the powers that have caused it.[13]

Descartes faces a problem more troublesome than he would like to admit. Time and again he insists that machines can *simulate* all the functions of life. But that is not the whole issue. The issue is whether in living things the *principles* of those functions are distinct from whatever principles might produce similar actions in nonliving things. The machine can, in fact, simulate, for an indefinite period, *every* action that a human being can perform, even those which, in the *Discours,* Descartes takes to be conclusive evidence of the presence of a soul. He too can distinguish those bodily actions which are caused by the soul only by reference to their principle, and not to any of their features.

The failure to confront directly the question of life in *L'Homme* evinces itself in two problems about life's beginnings. The first is that of the beginnings of life in individual animals, the problem of generation. In the *Treatise,* the

---

[11] An undated, but early, section of the *Primæ cogitationes circa generatio animalium* notes that "there are certain [things] common to all animals, like spontaneous movement, nourishment, etc.; those must come first in the consideration [of living things]" (*Primæ cog.,* AT 11:505). Descartes does not deny that living things *appear* to move spontaneously; indeed, if the words are construed correctly, he can affirm that they do. The principle of an animal's motion—which Descartes identifies with the heat of its blood—is indeed internal. But it is nothing other than a kind of motion in the blood, not a distinct entity like the Aristotelian soul.

[12] See Arriaga *De An.* d2§1.5n058, *Cursus* 641; Toletus *In de An.* 2c4q11, *Opera* 3:72va–b; and Des Chene 2000, §3.2.

[13] "I concede that it is not contradictory that the same thing, which occurs through vital action, should be able to be brought about by another agent without vital action, if it is done in a different way. For perhaps God can produce those entities or qualities of vital acts that are strictly speaking permanent without the concurrence of a power [in the tree], and then those actions would not be vital, even though the things or qualities produced would be the same" Suárez *De An.* 1c4n016, *Opera* 3:496–497. The parallel cannot be drawn very far. For Suárez, God can produce, in a tree, the actions of the vegetative soul, just as he can produce in any agent whatever actions its active powers can produce. Descartes, on the other hand, is supposing that in the *homme-machine* there is no agent at all.

men-machines are *given,* like actors on stage when the curtain rises. Descartes himself acknowledged the difficulty, and through much of his life tried to solve it. The *Description du corps humain* and the *Cogitationes circa generationem animalium,* some of which date from 1648, attempt to fill the gap. But Descartes never published them.

The second problem is that of the beginnings of animal kinds. Descartes can explain the origin of the planets, of the mountains and oceans of the earth, of vapors, exhalations, salts, and metals. But of animal kinds he says nothing. We can imagine God making each of them as he made Adam, from clay, but without breathing a soul into it. Not only is none of them born, but as a kind they originate from nothing. That there should be machines of almost infinite complexity in the world, and in bewildering variety—that too he cannot explain.

The first problem was too difficult. There were not enough years, enough funds, enough experiments, for Descartes to solve it. After considering his conception of the principle of life in Chapter 1, I examine in Chapter 2 his attempts, especially in the *Description du corps humain,* to devise a mechanistic theory of generation.

The second problem, when treated in a setting larger than that of natural philosophy alone, was trivial. In the seventeenth century creationism was not unscientific. Descartes conceded the truth of Genesis even when he had a convincing tale to tell.[14] Nevertheless, as we will see in Chapter 3, the givenness of plants and animals remains an undischarged teleological premise. Descartes can hardly avoid referring to the functions of their organs and operations. But there is no *locus* for ascribing ends to them except in God: yet his ends are inscrutable. The functional language that, like his opponents, Descartes uses to describe living things can be explicated only as a projection of human intentions onto a nature devoid of them, or of divine intentions that we are in no position to fathom. The *is like* of the fable of the world can be converted to an *is* only by forgetting the projection, a forgetting that elsewhere Descartes diagnoses as a primary error of common sense.[15]

Cartesian physiology met with sometimes vociferous opposition. The denial of souls, and so, it seemed, of feeling to animals, was the notorious sticking point. That aspect of the reception of Descartes' thought has been well

[14] The history of the earth can be explained naturally, or so Descartes believes. See *Discours* 5, AT 6:45; *PP* 3§45, AT 8/1:123–124.

[15] Descartes could have made up a story, using spontaneous generation, about the origin of species, consistent with the historical geology of the *Principles.* But there is no textual basis that I know of to show that he did, or thought he could, and I see no point in speculation. On projection, see To Regius, Dec 1641?, AT 3:455; To Élisabeth, 21 May 1643, AT 3:667.

studied. In this work I consider primarily the absence of any distinction in kind between living things and machines, and its implications for organic unity in the *bête-machine*. Not many seventeenth-century natural philosophers were persuaded that dogs and clocks differ only in complexity and originating cause. Fontenelle's riposte held good until this century, when finally the most mysterious of life's functions, that of reproduction, began to be chemically understood.[16]

Within a century of the writing of the *Traité de l'Homme,* and half a century after its publication, Leibniz, G. E. Stahl, and others had come to believe either that living things were animated by a special principle or else that matter was capable of far more than Descartes had granted. In the eighteenth century the "materialism" of La Mettrie and Diderot did not deny life to the living. On the contrary: it was inclined to see life in all things. It was not a physicalism, but a panvitalism, materialist only insofar as it denied the possibility of souls without matter. Yet it was not a simple revival of Aristotelian conceptions. Descartes divorced mind from life; the response of later philosophers was not to restore the Aristotelian tripartite soul, but to ascribe, as Leibniz did, *sentiments* or *petites perceptions* to all that lives—and thus to all material things.

---

[16] "Put a dog-machine and a bitch-machine near each other, and a third little machine may well result; but two watches will sit together all their life without making a third. And so we found, Madame B. and I, using our philosophy, that all things which, being two, are good at making themselves three, have a nobility much higher than machines do" ("Mettez une machine de chien et une machine de chienne l'une près de l'autre et il en pourra résulter une troisième petite machine, au lieu que deux montres seront l'une auprès de l'autre toute leur vie, sans jamais faire une troisième montre. Or, nous trouvons par notre philosophie, Mme B. et moi, que toutes les choses qui, étant deux ont la vertu de se faire trois, sont d'une noblesse bien élevée au-dessus de la machine") (Fontenelle *Œuvres* 1:31; quoted in Beaune 1980:9–10).

[1]

Self-Movers

Cartesian animals are self-moving machines, *automata* in the usual sense of the word.[1] To call them machines was not new.[2] The novelty was to combine the animal-machine with a new philosophy of nature, in which the actions of agents inferior to humans not only might but must be explained without reference to any "form" but extension or to any qualities but the modes of extension. Descartes had the formidable task of showing that the vegetative and sensitive powers of plants and animals are nothing other than the actions they exhibit by virtue of the "disposition" of their parts. He had also the less visible, but (as later events would show) crucial task of not allowing that demonstration to endanger the human soul. The machine must be powerful enough to

[1] Baker and Morris rightly point out that to render this as 'Animals are *mere* machines' is tendentious and to some extent anachronistic (Baker & Morris 1996:94; but see below §4.4, n. 32). Machines were marvels (see Daston & Park 1998:94–97). The fountain-works mentioned at the beginning of *L'Homme* (those at Saint-Germain-en-Laye, or the gardens built for Frederick V by Salomon de Caus at Heidelberg: see Bitbol-Hespériès in *Le Monde/L'Homme* 179n45, and *L'Homme*, AT 11:130–131) and the church organs to which Descartes also refers in *L'Homme* were the Difference Engines or *Great Westerns* of their day (see Beaune 1980:57–65). *Automaton* was sometimes used in its root sense to denote anything capable of self-movement. In that sense, the soul—and God—are automata but not machines.

[2] On machines and living things in antiquity, see Espinas 1903. Aristotle compares the vital spirits that carry heat out of the heart to "the instruments of art, by which artificial works are put together" ("Ideóque Aristoteles lib. 5. de Generatione animalium cap. 8. spiritum comparat artium instrumentis, per quæ artificiosa opera conficiuntur") (Coimbra *In Parv. nat.* "De vita et morte" 5, p87; the source is Aristotle *Gen. anim.* 789b10–13). John Cooper argues that the use of mechanical analogies in no way precludes—in Aristotle it rather invites—explanation by reference to ends (Cooper 1987:163–164). Gomez Pereira's work was mentioned to Descartes by Mersenne some years after the composition of *L'Homme*. Not untypically, he denied having seen it (To Mersenne 23 Jun 1641, AT 3:386).

make us believe that animals are machines, but not so powerful that it might be believed to *think*.

Mechanism is a *method*. It says yes to some modes of explanation and no to others. It promotes the analysis of capacities (see §4.2), and discourages the invocation of irreducible powers. It is also a substantive *doctrine* about bodies and change: bodies have only those properties that follow upon their being extended substances, and perhaps a few additional properties like impenetrability. As a doctrine, it invites us to regard natural things, if complex, as combinations of extended substances interacting only by way of collision. But it does not require us to do so. One may simply deny that animals fall within its scope. Conversely, one may compare animals or their organs to machines without adopting a mechanistic philosophy of nature. The machine is useful in comparisons because it is familiar, a glass or grid through which to grasp the less familiar. To serve that end, it need not be mechanistically conceived. A spring-powered clock may illustrate the storage of power whether one can explain springiness mechanistically or not. Aristotle, who was no mechanist, compares the parts of animals to simple machines like levers and bellows. In *De motu animalium,* for example, he argues that "the movement of animals is like that of automatic puppets," or *automata*.[3] Descartes, on the other hand, uses human-built machines, conceived *mechanistically,* to understand the actions of living bodies. The machine-model allows for the *transfer* to the living world, which at first is not understand mechanistically, of concepts that visibly do explain the actions of human-built machines.[4]

I will return to this point in Part II. For the moment I want to consider the explanation of self-movement by the scheme just sketched. Bemused or offended by the fantastic anatomy and the impossible hydrodynamics, or preoccupied with sensation and union, historians and philosophers often overlook Descartes' mechanization of the vegetative soul. He did not, indeed, achieve his aims; and many of the specifics of his physiology were rejected within a generation. But though it was, in that sense, a failure, it remains one of the most significant events in the history of physiology.

---

[3] Aristotle *De Motu anim.* 701b2–10 (trans. Nussbaum, 42). Aristotle immediately points out, however, that "in the puppets and carts no alteration takes place [. . . .] But in the animal the same part has the capacity to become both larger and smaller and to change its shape, as the parts expand because of heat and contract again because of cold, and alter" (701b10–16, trans. Nussbaum).

[4] Gaukroger's view of the analogy (as "just a means to an end: namely, a means for clearing the ground for a mechanist physiology," Gaukroger 1995:271) does not do justice to the role of machines in Descartes' work. The machine *is* a means, but Descartes extracts much more from the analogy than the subsumption of living things under the ontology of *Le Monde*.

1.1 Cycles

In Aristotelian definitions of 'life' self-movement is explicated both as *intrinsic* and as *immanent*.[5] The operations of life have sufficient proximate causes (efficient and final) within the agent, and many if not all of them terminate within the agent itself. Intrinsicness, it was agreed, is more general and more fitted to explain the commonsense distinction between living and nonliving things. A living thing has the immediate causes of its natural, characteristic changes within itself. The growth of a tree and the inflation of a balloon or the swelling of a wetted sponge resemble one another, but the balloon and the sponge increase their size only when acted on from outside. Moreover, the growth of trees perfects them: it is an action whose end is the agent itself that performs it.

The power of growth inheres in the organism itself; the food it eats is only the matter upon which the power acts, not the agency of growth. That power in turn has as its ground the form of the living thing—its soul, the spring from which flow all the powers characteristic of it as a kind of living thing. Anatomical structures of bone, flesh, and blood subserve its powers; but each power is distinct from whichever arrangement of those substances is fitted to be its instrument in this or that species. It is to the powers first of all, and only through them to the structures, that the actions of living things are referred.

God can, nevertheless, bring about in living things any of the actions that naturally occur only by their powers.[6] A world might exist, indistinguishable from ours, in which nothing lived, because no action of a living thing was intrinsic. But that is irrelevant to giving the causes of actions here and now. Though the Aristotelian does not insist on the point as Descartes would, the impossibility of our being wholly deceived about the efficacy of second causes is understood through all of Aristotelian natural philosophy. Arguments merely reinforce that basic tenet. That "created agents truly and properly bring about

---

[5] See Des Chene 2000, §3.2, for a discussion of some Aristotelian questions on the definition of life.

[6] Whatever God brings about through secondary causes, he can bring about immediately. There can be no *essential* instruments of divine power. Even my judgments can be brought about, not by my intellect and will, but immediately by God. A body could be made to say *exactly* what the body of the living Descartes says, for as long as you please. The criterion of *Discours* 5 (AT 6:56–59) will not save us from that possibility (cf. Baker & Morris 1996:89). Whether animals have souls is not a matter of whether their actions and those of automata could be exactly similar. It is a matter of what they are made of: for Descartes, they are made of *res extensa,* from which any quality or form like those in Aristotelian physiology is excluded.

the effects which are connatural and proportionate to them" is proved, first of all, by experience:

> for what is better known by sense than that the sun illuminates, fire heats, water cools? If they should say that although we do experience those effects being brought about when those things are present, still they are not brought about [by those things], clearly they destroy all the force of philosophical argument, because we cannot experience in any other way the emanation of effects from causes or grasp causes by effects.[7]

The argument applies equally to living things. It would be as much a mistake to suppose that trees do not really grow (where the subject-verb phrase is understood on the model of "the sun heats," "the philosopher thinks," that is, as expressing an action, and not, say, pure becoming), or that animals do not really nourish themselves.[8] Yet that is what Descartes asks his audience to believe. The animal-machine does not grow; matter accretes to it. Nor does it see; rather light and the machine interact in a certain way.

The language of agency is ill-suited to Cartesian nature generally. Descartes construes the Aristotelian explanation of the fall of a stone as an intentional explanation: 'the stone falls (i.e., to reach the center of the world)' parallels 'the woman walks (i.e., to reach her house).' His own explanation, in terms of pressure, and thus of collisions, must rest on equations derived from the laws of motion. Though not much used by Descartes—and never in his physiology—, the language of equations offers no hold for agency. So too with animals. 'The dog walks' only superficially resembles 'the philosopher walks.' If we could express the motions of the parts of the dog through time in a set of equations, *that* would express, without erroneous projections of agency, the event of the dog walking. But when we say that the philosopher walks home, then language does not mislead us; the equation would, since it fails to express the fact that the philosopher's walking home is an *act*.

Aristotelian arguments on behalf of the soul—on behalf, that is, of the claim that the principle of vital operations is the substantial form of a living

---

[7] Suárez *Disp.* 18§1n06, *Opera* 25:594.

[8] In a slightly different context (arguing that the soul is not a *forma assistens*) Suárez writes: "motions, which are brought about by a substance assisting [*assistente*] another, in which it acts [*assistit*], cannot be vital actions of the substance in which the other is acting: an assisting substance, therefore, does not produce life in such a body, and is therefore not its soul. [. . .] Hence the heavens are not said to move [*se movere*], but to be moved by another, which cannot be said of living things. For plants truly nourish themselves [*se nutrit*], and the horse sees, and so on" (Suárez *De An.* 1c2n013, *Opera* 3:471).

thing—fall into two main sorts.[9] The first are specializations of arguments in natural philosophy on behalf of substantial forms generally: there must be *one* underlying principle in which all the powers of a thing are united, accidents are instruments of substances, and so forth. The second are arguments to the effect that primary qualities are insufficient to bring about the higher operations of the soul, especially knowing and loving in humans.

Descartes' strategy is to show that no such form is needed. It succeeds better against the second sort of argument than the first, though even then generation resists explanation. It yields, moreover, only what an engineer would call a "proof of concept," a proof designed to show that an automaton *could* be made to perform all the required actions in the appropriate circumstances.[10] We must take Descartes' word for it that when swallows return in the spring, "they act [. . .] in the manner of clocks" (To Newcastle 23 Nov 1646, AT 4:575), and that

> everything that honeybees do is of the same nature, and the order that cranes observe in flying, and [the order of battle] which is observed by apes when they fight, if it is true that they observe some order, and finally the instinct by which they bury their dead, is no more strange that the instinct of dogs and cats, who scrape the ground to bury their excrements, even though they almost never bury them: which shows that they do so only by instinct, without thinking about it. (Ibid. 575–576)

Anyone who has seen a cat scrape a bare floor can testify to the occasional lack of fit between act and end. Yet the very fact that we see it as failing to achieve that end suggests that ends are normally invoked in explaining the acts of animals. With tornados, on the other hand, even if they seem to hit trailer parks with unusual frequency, we are not tempted to attribute failure when a tor-

---

[9] See Des Chene 2000, §4.2, for discussion and references.

[10] Gueroult notes that "the possibility of fabricating artificially machines that would imitate life, even perfectly, does not permit one to conclude, as Descartes sees it, that organisms are identical to machines; it authorizes one only to consider that such a conclusion is simply conceivable." In order to prove it with certainty, one would have to show "that *no other mode of explanation is possible*." (Gueroult 1970:). This is correct, as is Gueroult's subsequent argument that the grounds for denying souls to animals are the mutual exclusion of mental and physical properties in the same substance, and the denial of *reason* to animals (see the letter to Newcastle cited above). Nevertheless, even though it cannot be asserted that the absence of souls in animals (or the presence of souls in other humans) is *metaphysically* certain, still the claim that animals are machines is *morally* certain by the criterion of *PP* 4§205. The Coimbran commentary on *De Anima* likewise acknowledges that we cannot prove that oysters and other lowly creatures lack a rational soul; we can only argue that there is no evidence for it, and that in bodies having such primitive organs it could not exert its powers, and would therefore be useless.

nado misses its "target." For Descartes, tornados and cats are of a piece. The ascription of ends is so closely tied to the ascription of thoughts that animals can no more act toward ends than they can think.[11] If the actions we attribute to "instinct" are indeed thoughtless, matter and its properties will suffice to explain them—even in us.

The first step in the proof of concept is to show that actions attributed to the vegetative part of the soul can be performed by the man-machine, which by hypothesis lacks all properties but those it can have by virtue of being composed of *res extensæ*. After passing quickly over the gross anatomy of the *homme-machine,* Descartes begins with digestion.

> First of all, foodstuffs [*les viandes*] are digested in the stomach of this machine, by the force of certain liquors, which, sliding in among their parts, separate them, agitate them, and heat them: just as ordinary water acts on [the parts of] quick-lime, or *aqua fortis* on those of metals. [. . .]
>
> And know that the agitation that the little parts of those foodstuffs undergo in being heated, joined with that of the stomach and the intestines that contain them, and by way of the disposition of the little filaments of which the intestines are composed, brings it about that to the extent that they are digested, they de-scend little by little toward the conduit through which the largest of them must leave; meanwhile the more subtle and the more agitated encounter here and there an infinity of little holes, through which they flow into the branches of a large vein that carries them toward the liver, and into other [veins] that carry them elsewhere, without there being anything but the smallness of the holes that separates them from the larger [particles. . . .]
>
> [. . .] It should be noted also that the pores of the liver are so disposed that when this liquor enters into it, the liquor is subtilized, elaborated, takes on color, and acquires there the form of blood. (AT 11 : 121–123)

Until the last sentence, with its 'subtilize' and 'elaborate,' and its mention of color, the description does hold superficially with the promise of a strict Cartesian mechanism. The only qualities distinguishing the particles of food

---

[11] In his reply to the Sixth *Objections,* Descartes writes that "it appears that this idea of heaviness was drawn from the idea I had of mind principally because I took heaviness to carry the body toward the center of the earth, as if it contained in itself some knowledge of it [i.e., the center of the earth]" (*Med.* AT 7 : 442; see also To Élisabeth 21 May 1643, AT 3 : 667; To Arnauld, July 1648, AT 5 : 222–223; Garber 1993 : 96–102). Descartes treats all *directedness* toward ends as *intentionality.* There can be no intentional relations except between thoughts and things, and thus no directedness toward ends except if those ends are conceived by a mind.

that are eventually excreted from those that go on to the liver are size and "agitation." Heat, we have learned in *Le Monde,* is the violent motion of little parts of matter. Color is explained only in the *Météores,* published with the *Discours* in 1637; but Descartes was undoubtedly drawing on work done before. Descartes offers no explanation of the redness of blood, nor of how the chyle is made red. The reader must grant that somehow this can be dealt with.[12]

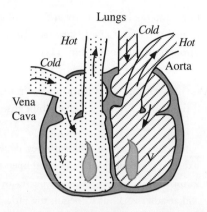

Figure 1: Cardiac circulation. Dotted areas are venous blood, striped areas arterial. V = ventricle.

Digestion is the first step in a cycle of replenishment of the blood. The next step in the cycle is the circulation of the blood, which passes via the liver to the heart.[13]

Know that the flesh of the heart contains in its pores one of those fires without light that I have spoken of earlier, which makes it so hot and ardent that as the blood enters one of its two chambers or concavities, it quickly inflates and dilates: just as you can experience with the blood or milk of some animal, if you pour it drop by drop into a very hot vase. And the fire in the heart of the machine that I am describing to you, serves for nothing else but thus to expand, heat, and subtilize the blood, which falls continually drop by drop through a tube of the *vena cava* into the concavity of its right side, from which it breathes

[12] In *Le Monde* Descartes shows that there are just three elements, distinguished by size, shape, and characteristic motion: the third element, sometimes called "earth," consisting in fairly large bits of extended stuff; the second element, consisting in much smaller particles, spherical in shape; and the first, yet smaller, very rapidly moving particles that fill the interstices between the other two. Light is the motion or tendency to move of second-element particles. In the *Description,* Descartes, applying the theory of color from the *Météores,* holds that the blood is red because its parts are so ramified and tangled that in the spaces between them only first-element particles are found. Those particles, moving hither and thither with their usual great speed, cause second-element particles striking the surface of blood to turn rapidly relative to their speed, which is the complex property of light that yields sensations of red (*Descrip.* AT 11:255–256; *Météores* 8, AT 6:333–334).

[13] Louis La Forge, in his notes to the first edition of *L'Homme,* corrects Descartes on this point: the chyle does not, for the most part, go straight to the liver. Descartes, he notes, would have known this after *L'Homme* was written, because he was familiar with the discoveries of Asellius and Pecquet (*L'Homme* AT 11:122, quoting *De Hom.* 180–181).

forth into the lungs, and from the vein of the lungs, which the Anatomists have named the *venous artery,* into its other concavity, from which it is distributed throughout the body. (Ibid. 123)

After describing the pulmonary circulation and the function of respiration and the lungs, Descartes gives the cause of the pulse.

The pulse, or beating of the arteries, depends on eleven little membranes which, like so many little doors, close and open the entries of the four vessels that open into the two concavities of the heart; for at the moment when one beat ceases, and another is about to begin, the little doors at the entries of the two arteries are exactly shut, and those at the entries of the two veins are open: so that two drops of blood will not fail to drop through the two veins, one in each concavity of the heart. Then, as these drops of blood are rarefied, and extend all at once into a space incomparably greater than the space they have occupied heretofore, they push on and close the little doors at the entries of the two veins, and prevent more blood from entering the heart, and push on and open the [doors] of the two arteries, through which [the expanded drops] enter immediately and with force, causing the heart and all the arteries of the body to inflate at once. But shortly thereafter, the rarefied blood condenses again or penetrates the other parts [of the body]; and so the heart and the arteries deflate, the little doors at the entries of the two arteries close again, and those at the entries of the two veins open again, and make way for two more drops of blood, which will again inflate the heart and the arteries.[14]

The description is, banally, causal. Each event is the proximate cause of the next. The reader, however, is more likely to be persuaded by the coherence of the narrative, and the comparisons drawn along the way, than by any implicit application of the laws of motion.[15] The previous passage has suggested an *expérience* by which to confirm the one element that might not be me-

---

[14] See also *Discours* 5, AT 6:49–50; *Descrip.* AT 11:231–232; To Beverwick 5 Jul 1643, AT 4:4–6.

[15] Nevertheless that is supposed to underlie not only the explanation, but the comparisons. "The construction of those [arterial] valves is such that necessarily, according to the laws of Mechanics, by the sheer impetus of the blood they are opened" ("Ea enim est fabrica istarum valvularum, ut necessario, iuxta leges Machanicæ, ex hoc solo sanguinis impetu hæ aperiantur") (To Beverwick 5 Jul 1643, AT 4:5). "All these things [in the explanation just given of the pulse] are mechanical, just as the experiments too are mechanical, by which it is proved that there are various anastomatoses of the veins and arteries, through which the blood flows out of them" (ibid. 5–6). See also To Mersenne 20 Feb 1639, AT 2:525 and *Descrip.* §66, AT 11:277, quoted below.

chanically obvious—the sudden ebullition and rarefaction of the blood when it enters the heart after being cooled and thickened in the lungs.[16]

Descartes has not merely explained how the heart beats *once*. He has explained the *cycle* of expansion and contraction. Harvey, whom Descartes generously acknowledged, proved that the blood of the arteries returns through the veins to the heart in a "perpetual circulation."[17] But he had no explanation of the regular, repeated exertion of the heart's "pulsific force" (*vis pulsifica*). Subsequent work has shown that Descartes' explanation is mistaken. The ebullition of the blood, as Harvey already believed, is an invention. Harvey, insisting that the blood is forced from the ventricles by periodic muscular contractions, was closer to the truth, though no closer to an explanation of the heart's regular motion.[18]

In *L'Homme* and subsequent texts on the heart, one basic feature of the self-motion of animals is accounted for. Self-motion—self-*perpetuating* motion—is a motion that, in exhausting itself as the blood condenses in the arteries and the heart deflates, creates the conditions that cause it. Thus is overcome one hurdle on the way to a mechanical explanation of the functions of the body. Even though clocks, for example, produce periodic motions, they do so in a way that has no obvious counterpart in organisms. Nature, in the words of Leonardo, "does not use counterweights"—or springs.[19] In the cycle of the blood, the motor is the heat of the heart, which makes the blood expand in the ventricles.

The next cycle is the making of blood. In digestion, we have seen, blood is formed in the liver and flows to the heart. But why must it be formed, if it perpetually circulates through the body?

[16] Descartes also likens the process to fermentation. Just as the making of bread dough requires a "starter," in the heart a small portion of blood, "which fills the inmost recesses of its ventricles, acquires a new degree of heat and a certain nature as of yeast" (To Plempius 15 Feb 1638, AT 1:530; cf. 523).

[17] In the *Discours,* Descartes mentions "a physician from England, to whom one must give praise for having broken the ice" on the question of the circulation of the blood (*Discours* 5, AT 6:50). See also To Boswell 1646?, AT 4:700). His first response was more guarded: "I have seen the book *De motu cordis* which you once mentioned, and found that I differed a little from its opinion, although I saw it only after having finished writing on the matter [in the *Traité de l'Homme*]" (To Mersenne Nov or Dec 1632, AT 1:263).

[18] Harvey, supposing that the heat of the blood is the principle of life, hints that the influx of blood into the coronary arteries sets the stage for the next contraction of the heart. See Pichot 1993:302. On the Harvey-Descartes controversy, see Gilson 1984, Bitbol-Hespériès 1990, Fuchs 1992, Grene 1993.

[19] "Although human ingenuity in various inventions with different instruments yields the same end, it will never devise an invention either more beautiful, easier, or more rapidly than does Nature, because in her inventions nothing is lacking and nothing is superfluous, and she does not use counterweights, but places there the soul, the composer of the body." (Galluzzi 1987:260, quoting Leonardo da Vinci *Anat. stud.* f114r = Royal Library 19115r).

the same blood passes and passes again several times, from the *vena cava* into the right cavity of the heart, then from there by the arterious vein into the venous artery, and from the venous artery into the left cavity, and from there through the great artery into the *vena cava:* and so a perpetual circular movement is brought about, which would suffice to sustain the life of animals, without their having any need to drink or eat, if none of the parts of the blood went outside the arteries or veins as it flowed in this way; but many parts go out continually, and the loss is made up by the juice from foodstuffs, which comes from the stomach and the intestines. (*Descrip.* AT 11:239)

Blood is used up in growth, which is the accretion of blood particles onto parts of the body:

if it is the body of a child that our machine represents, its matter will be so tender, and its pores so easy to enlarge, that the parts of the blood that enter into the composition of the solid members [of the body] will be ordinarily a little larger than those they take the place of; or it may even happen that two or three succeed upon one alone, and this will be the cause of growth. (*L'Homme* AT 11:126)

Other particles are used to form digestive fluids and saliva, or excreted as urine and sweat. A few, "the liveliest, the strongest, and the most subtle," move up into the brain, there to form the animal spirits. The next best descend to the organs of generation, to make seed or a new machine (*L'Homme* AT 11:128).

The "interior sensation" (*sentiment intérieur*) of hunger occurs when the digestive fluids, not finding enough food "to occupy all their force in dissolving it, turn against the stomach itself," and agitate the nerve-endings there, so as to cause "hunger" in the machine, and hunger in a soul joined to the machine. The satisfaction of "hunger"—the functional state in which the gastric nerve-endings are being agitated—refills the stomach. The digestive fluids, the *aquæ fortes* that break up the food, find a better target than the stomach itself, and the nerve-endings cease to be agitated in that fashion.

Before it can circulate into the body, the blood must be cooled and thickened by the lungs, which, like the heart, but more slowly, expand and contract periodically. The explanation of this cycle is obscure, especially since the figures printed with *L'Homme* in 1664 and in the Latin edition of 1662 are hard to make out, and differ among themselves.[20] The system of nerves

---

[20] This even though two of the three are said to be from Descartes' own sketch. Or perhaps that is the problem—he admitted he was no draftsman. Of the figures in his own manuscript copy of the *Essais,* he writes that they are traced by his hand, "c'est à dire très-mal" (To

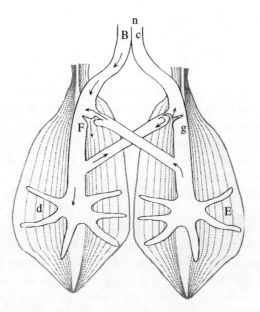

Figure 2: Antagonistic muscles (based on AT 11: 134–135 and Fig. 3). The tubes *BF* and *cg* are nerves; *d* and *E* are muscles, *F* and *g* are valves. The arrows in the nerves indicate the flow of animal spirits when *d* is expanding and *E* is contracting. Valve *F* is being pushed down by the flow from *B* to *F*, and valve *g* is being pushed from the left by the pressure from *d*. At the top *n* is a channel supplying both *BF* and *cg*.

controlling a pair of antagonistic muscles resembles what we would call a flip-flop.

We are to imagine (see Fig. 2) that at the outset the animal spirits are moving down the nerve-tube *BF*, filling up the muscle *d* on the left. The pressure of the flow pushes the valve *F* downward (it should be imagined as a kind of piston; when it is all the way up it blocks the channel coming up from *E*). Meanwhile backflow toward *g* pushes that valve upward, closing off the diagonal tube out of *d*. The muscle *d*, therefore, not only fills up, but depletes *E* in doing so, as the arrows indicate.

Mersenne, Mars 1636, AT 1:339; see also To Huygens, 5 Oct 1637, AT 1:447). The figures differ especially in the crucial detail of the valves *F* and *g*. It leads one to wonder if the physical puzzle of making *F* *open* when the spirits are pushing on it from *B*, but *closed* when spirits are flowing up the diagonal channel from *E*, a behavior that would seem to require inconsistent linkages, did not lead the illustrators simply to follow their fantasy. The crudest of the three versions (from 1662) is simply unintelligible. The other two do better, but I don't think either would work. In Figure 2 I have done my best, representing the valves not as swivelling but as moving up and down, like the valves in an automobile engine.

there are certain membranes around the muscle d, which press more and more as it inflates, and are so disposed that before all the spirits in muscle E have passed toward [muscle d], they stop their flow, and make them overflow through the tube BF, so that [the spirits] in the channel n change their course; and by that means, being directed into the tube cg, which at the same time is opened by them, they make the muscle E inflate, and deflate the muscle d; this they continue to do as long as the impetus [*impétuosité*] with which the spirits contained in the muscle d, compressed by the membranes around it, tends to exit [from d]. Then, when this impetus has no more force, they follow again of their own accord [*d'eux-mêmes*] their course through the tube BF; and thus never cease to inflate and deflate the two muscles.[21]

There are, I must say, easier ways to produce the same behavior. One can only lament Descartes' lack of *mechanical*, as opposed to *geometric*, intuition. Even a brief look at the anatomical drawings of Leonardo shows that in them a mind of a different cast is at work, an eye (and hand!) that can see how a linkage will operate. But even though Descartes falls short as an engineer, he has the right idea, an idea perhaps unprecedented in physiology.[22] The system—all fictitious, alas—of nerves, with their spirit flows and valve-modulators, is a system that controls its own operations.[23] Descartes' insight was to understand—however tenebrous the details may be—that self-regulation is within the capacity of a machine.

It would be an error, all the same, to suppose that Descartes has anticipated the notion of negative feedback. While it is true that the overflow of spirits from muscle d eventually influences the flow of spirits into d, a crucial element is missing: namely, a *dedicated* link l which, receiving input from a component C, controls the behavior of a second component B causally prior to C. A close examination of Figure 2 will show that the flow into d, though it controls the flow out of E, is not linked backward; it shuts itself off simply by filling d, and forcing the flow from n at the top to be diverted to the tube cg.

---

[21] *L'Homme* AT 11:139. The *Passions* contains a vaguer description of the same setup (*PA* art11, AT 11:335–336).

[22] There is some evidence that Descartes was proud of this part of his theory. Recounting the history of the manuscript of *L'Homme*, and the copies he has allowed to be made, Descartes laments the fact that Regius has somehow got hold of one, and taken from it "that nice piece on the movement of the muscles" ("cete belle piece du mouuement des muscles") (To Clerselier 23 Nov 1646; AT 4:567).

[23] On the hollowness of the nerves, see Kepler *Ad Vitell.* (1604). No doubt the valves in the nerves are proposed by analogy with the newly discovered valves in the veins, announced by Fabricius ab Aquapendente *De ven. ost.* (1603). The discovery was included in Bauhin's *Theatrum anatomicum*, which Descartes consulted (see Bitbol-Hespériès 1996:180n58).

This is no more a feedback loop than is a cup running over and spilling onto the floor. Descartes has, nevertheless, constructed a mechanism that, operating in a fixed pattern, eventually creates the conditions under which the pattern can begin anew. That is the keystone of the argument which allows him to banish the vegetative soul.

The depletion and formation of the blood is a second, larger, cycle that includes the circulation of the blood, digestion, and respiration. One further cycle is necessary to complete the explanation of the self-maintenance of the machine. The sense of taste, first of all, favors aliment that is good for us (or so Descartes says): "the same little parts of food which in the mouth are able to enter the pores of the tongue" are those that in the stomach can pass into the blood; and "only those that tickle the tongue moderately, and which may thereby make the soul sense an agreeable taste, are entirely proper" for digestion (*L'Homme* AT 11:146). Much later, discussing the determination of the animal spirits by the pineal gland, Descartes supplies the last piece of the puzzle. When the movements that give rise to hunger occur, and "when nothing is presented to any of the senses, nor to memory, that seems proper for eating," the spirits caused by the condition of the stomach "will betake themselves to a location where they will find several pores disposed to conduct them indifferently into all the nerves that can serve for seeking or pursuing an object" (195). The explanation is vagueness itself. But what is clear is that Descartes believes his machine can be made not only to replenish its blood when it has food, but to seek food when it does not.

The machine, then, is a self-mover, first of all, because some of its movements are comprised in self-perpetuating cycles: the periodic contraction and expansion of the heart, the depletion and replenishment of the blood, hunger and satiety, desiring food, seeking and obtaining it. We see here one avenue for the development in the machine of autonomous "desires," which lead the machine to act so as to maintain itself.[24]

1.2 The *feu sans lumière*

Descartes has so far accounted for two of the three basic operations ascribed to powers of the vegetative soul: nutrition and growth. Setting aside the third, generation, for Chapter 2, I consider here the problem of the "principle of

---

[24] A second avenue, not part of a cycle, and requiring the intervention of the senses, is the spontaneous withdrawal of the machine from noxious stimuli: in Descartes' example, the automaton removing its hand from near a flame because of the pain of the heat (*L'Homme* AT 11:191–193; see Canguilhem 1977. c.2).

life."[25] At the end of *L'Homme,* Descartes, summarizing his achievements, lists all the operations he has explained, and concludes that

> it is not necessary to conceive [. . .] in [the machine] any other vegetative or sensitive Soul, nor any other principle of movement and life, other than its blood and its spirits, agitated by the heat of the fire that burns continually into its heart, and which is of no nature other than all the fires that are found in inanimate bodies. (*L'Homme* AT 11:202; cf. *PA* art8, AT 11:333)

Fire is the only principle that functions as a *mover.* The others merely convey coherent motion and chaotic heat. Hence it is reasonable to call the fire of the heart and its heat *the* principle of life, as Bitbol-Hespériès has argued.

In the second chapter of *Le Monde* we learn that fire is composed of "small parts that move separately one from another, with a very quick and active movement, and which, having that sort of movement, push and move with themselves the parts of the bodies they touch that do not offer too much resistance."[26] Like air, it is a fluid, but its parts are larger; that is why air does not consume things but fire does.[27] The fire in the heart, however, is a "fire without light" (*feu sans lumière*).[28] That fire, we learn in the *Principles,* occurs in the stems and leaves of new-cut hay.[29] After a plant is cut and is beginning to dry out, some of the pores that contain sap will become narrower than usual, so much so that the sap, instead of being surrounded as it usually is by particles of both the first and second elements, is surrounded and follows the very rapid course of motion of first-element particles alone; and "when the particles of sap thus follow the course of the first element, they have much more force to impel bodies they encounter than would the first element

[25] Bitbol-Hespériès 1990 is the fundamental text; see also the summary in Bitbol-Hespériès 1988 and the notes in *Le Monde/L'Homme,* 175n22, 176n23.

[26] Compare *Descrip.* §72, AT 11:280–281: "I recognize no other fire or heat in the heart except the agitation of the particles of the blood."

[27] *Le Monde* AT 11:8, 15; see *Discours* 5, AT 6:44, 46; *Météores* 1, AT 6:236; *PP* 4§80, AT 8/1:249–250.

[28] *L'Homme* AT 11:123.

[29] Descartes more than once compares the process by which the blood is heated in the heart with fermentation and the spontaneous heating of rotting vegetation. See, for example, To Newcastle Apr 1645?, AT 4:189. There is an explicit reference to the *Principles* near the end of the *Description:* "the movement of the diastole has been caused from the beginning by heat, or the action of fire, which, according to my explanation in the *Principles,* could not have consisted in anything else but that the matter of the first Element, chasing out [the matter of] the second from around some parts of the seed, has communicated its agitation to them." (*Descrip.* §72, AT 11:280–281).

alone, just as one sees that a boat following the course of a river has more [force] than the water in the river."[30] The example leads to a generalization:

> whenever some hard body is heated by the admixture of some liquid, I hold that this occurs because many of its pores are of such a size that the particles of the liquid admitted into them are surrounded by matter of the first element alone. And for a similar reason, I hold, one liquid poured into another [may heat it up], since always one or the other is constituted by branched particles, somehow hooked into and connected with one another, so that it can fill the office of a solid body. (*PP* 4§93, AT 8/1:257–8)[31]

The same story explains how in shooting stars and thunderbolts fire *with* light is produced (*PP* 4§88–89, AT 8/1:253–4).[32] The difference, as far as I can see, is that when a meteor falls, the fire in it can communicate with second-element particles in the surrounding air, but in rotting hay and in the heart it cannot. Only in love poems does the heart glow: the real heart is a heart of darkness.

One feature of the explanation is paramount: fire is mechanical, and fermentation, traditionally a chemical process, is a kind of combustion. The head on a mug of beer and the foam atop boiling milk are brought about in the same way. Descartes will preserve that view by whatever arguments are necessary to save the phenomena. Plempius, for example, objects that in fish

[30] *PP* 4§80, AT 8/1:257, 9/2:251. I follow the French version, which differs from the Latin mostly in being more explicit.

[31] The *feu sans lumière* has nothing to do with air. In the body, the lungs alter the blood, and thereby indirectly help nourish the fire in the heart. But air itself has no role in combustion. According to the Coimbrans, when we breath in, the lungs draw out from the heart the vapors that result from the coction of the blood in the left ventricle, which are expelled when we breathe out (*De respiratione, In Parv. nat.* 58–59; see Gilson 1984:61). Those who think degrees of verisimilitude are totally ordered might want to test their measures on the Coimbrans' theory and that of Descartes.

[32] In taking the fire in the heart to be a *feu sans lumière* Descartes opposes himself, as Bitbol-Hespériès argues, not only to Aristotle and Galen, but to Kepler. In the *Paralipomena ad Vitellonem* Kepler compares the heart with the sun, each being the source of heat and life in its system. The heat of animate things "penetrates into the arteries through the heart, in which resides, according to Fernel, something resembling a permanent flame; for my part I make bold to affirm that a genuine flame exists there." Indeed "the very form of attraction and expulsion, the machinery of nature in the valves, proclaim with a powerful voice that the heart contains a working flame in order to sustain life by the contributions of the elements that enter the heart, and to expel by the same path the remains and the proper productions of this flame" (Kepler *Ad Vitell.* 1prop32, quoted in Bitbol-Hespériès 1990:73–74). But Descartes, though like Kepler he "opts for a true fire," not distinct in kind from the fire in inanimate things, nevertheless removes from it "one of the sensible qualities of the Keplerian cardiac fire: light" (Bitbol-Hespériès 1990:74).

the heat of the body is "paltry, more like cold [than heat]; but their heart beats as quickly as ours."[33] Descartes first replies that "even if in fishes [the heat] does not feel intense, still in their heart it is much greater than in any other member."[34] Even so, says Plempius, that heat is "not so great that it can rarefy the blood, let alone so quickly." After all, our hands are hotter than a fish's heart, but fish blood does not boil on them.[35]

Descartes does not answer the objection directly. Instead he describes experiments with the "tiny heart" of an eel, the gist of which is to show that heat and the flow of blood are both necessary for the pulsing of the heart. An eel's heart excised in the morning, he says, had begun to dry up some hours later, but on being warmed just a little it began to beat. In another experiment, he put an eel's heart "into the same eel's blood, which I had saved for this use, and I made it heat up, so that it beat no less quickly and markedly than in the living animal."[36] As if that weren't enough, he adds that the blood of the eel differs greatly from that of warm-blooded animals. The "most subtle parts" of human or dog blood "fly out, and what remains turns partly into water, partly to a lump of earth." But the eel's blood "remained, I do not say uncorrupted, but at least, so far as could be seen, unchanged, and many vapors issued from it, so much that they, if heated even a little, rose up like densest smoke" (67).

Countering Harvey's experiments, which purport to show that the blood exits the heart because it is squeezed out, Descartes suggests that "all the same effects might proceed from another cause, namely, the dilation of the blood" (*Descrip.* AT 11:239). After describing two experiments that favor dilatation, Descartes notes that since the blood is changed in the heart (as Harvey too believes) Harvey requires at least two principles where Descartes requires but one:

> Not only must one imagine some faculty that causes this movement, [a faculty] whose nature is much more difficult to conceive than anything he claims to explain by it; but also one must suppose other faculties that would change the qualities of the blood while it is in the heart. Instead, by considering only the dilation of the blood, which necessarily must follow from heat, a heat that everyone recognizes to be greater in the heart than in any other part of the body: one sees clearly that dilation alone suffices to move the heart in the manner I have described, and at the same time to change the nature of the blood as much as ex-

---

[33] Plempius to Descartes Jan 1638, AT 1:496. On their exchange, see Grene 1993.
[34] Descartes to Plempius 15 Feb 1638, AT 1:529. This was a traditional view.
[35] Plempius to Descartes Mar 1638, AT 2:53.
[36] Descartes to Plempius 23 Mar 1638, AT 2:66.

perience shows it to change; and even as much as one may imagine it must be changed, in order that it should be prepared, and made more fit to serve as nourishment for all the members [of the body] and for all the other uses it has in the body. (*Descrip.* AT 11:243–244)

Harvey holds that the heart has the power of contracting, *and* that in it the blood is heated. But Descartes allows just one "faculty" in the heart, and that not a mysterious power, but only the mechanically explicable heating of the blood. The heart itself is entirely passive. Its only "power" is that of resisting expansion, like a bladder filled with air. The fibers of which it, like all organs, is composed, tighten as its cavities expand. That, and not the operation of a contractile faculty, is why the heart hardens as the blood rushes forth from it. The phenomena observed by Harvey may indeed have a cause other than the one he puts forward.[37]

Descartes was far from presenting a purely a priori deduction of the structure and operation of the heart.[38] As Marjorie Grene observes, he "seems positively to glory in the evidence of the senses, in what can be seen and touched."[39] But the evidence Descartes brings forward, detailed and abundant though it is, he marshals to confirm, or save, the principles of his natural philosophy. There is not (in fact), nor can there be (in theory), any quality or form peculiar to living things. Descartes agrees with many of his predecessors in holding that the principle of life is heat, the heat of the heart; he takes over many claims from older physiologies—that the heart is the hottest organ, for example. Harvey breaks more firmly with tradition on such matters. But Descartes admits those claims only if they can be interpreted mechanistically: ideally, in terms of the laws of motion, but practically, by way of comparisons with nonliving systems whose mechanical nature he takes to be obvious.[40]

[37] On Harvey's response to Descartes, see Gilson 1984:95–101, which includes an excerpt from the *Exercitationes*.

[38] On method in Harvey and Descartes, see Grene 1993. Wear 1983 is a lucid presentation of Harvey's methods.

[39] Descartes vaunts himself for having made an exercise of dissection, to the point that he believes that "hardly any physician has looked as closely as I have" (To Mersenne 20 Feb 1639, AT 2:525). However that may have been, it is clear that Harvey was the more careful, the more disciplined observer. All three experiments adduced by Descartes against Harvey's theory of the operations of the heart are erroneous (Grene 1993:335).

[40] Descartes seems to have made no attempt to verify his hypotheses directly by constructing hydraulic machines with properties analogous to those of the heart, the vessels, and so forth. In the first half of the eighteenth century, the construction of such machines was undertaken, though unsuccessfully. François Quesnay writes that "The principles and proofs I found [in his "project" of a scientific physiology] seemed to me to be very much in agreement with the laws of hydrostatics and the disposition of the organs [. . . .] I had constructed for me a hydraulic machine to convince myself entirely of the facts I had discovered" (Quesnay *Observations*, "Préface"). He built branching iron tubes to study the flow of blood through the arteries (Quesnay

There are no organic substances, no vital powers, only the three elements of *Le Monde* and their various configurations and motions. Life must disappear if, as Descartes believes, there is but one matter, and that matter is *res extensa*.

In Aristotelian natural philosophy, and medical works of an Aristotelian cast, the soul, of course, is the principle of life. For Suárez, this is even a matter of definition. Because, as we have seen, no accident can act except as the instrument of some substance, animal heat alone cannot be the principle of life. Only a substantial form, the soul, can be. Unlike some of their contemporaries, Suárez and Toletus show no tendency to mysticize vital heat or the cardiac fire. The heat of animal bodies is produced by processes akin to fermentation. Whatever Aristotle may say, it is not different in kind from elemental heat.[41] Like Descartes', the physiology of the Aristotelians is naturalist. But unlike his, it is "animist," it attributes to every living thing souls which in their vegetative and sensitive parts resemble ours.

Descartes can account for the *appearance* of self-movement, our at first overwhelming impression that animals have an internal principle of motion, a soul resembling our own. The animal-machine, once started, will *continue* to move in just the ways we experience. Each of its cycles needs to be "initialized," but once that is done, its blood will circulate, it will periodically replenish and thicken its blood by eating, digesting, and breathing, and will do what it should in order to eat, not because continued existence is the end toward which it acts, but because a *prior* state of hunger naturally causes it to seek food. Purely local efficient-causal interactions suffice, or so Descartes intends to persuade us, to explain the cycles that an Aristotelian would explain in terms of the soul, its powers, its natural actions, and its ends.

Traité 160–161; see Doyon & Liaigre 1957:296–297). Vaucanson, best known for his mechanical duck and flute-player (Diderot *Encyclopédie* s.v. "Automate"), presented his project of "anatomies mouvantes" to the Académie des Sciences et Beaux-Arts at Lyon in 1741. A transcript of the session says that "he claims that one will be able to perform experiments, using this automaton, on the animal functions and arrive at inductions by which one may know the different states of health in men so as to remedy their ills" (Doyon & Liaigre 1957:298). Another physician, Claude-Nicolas Le Cat, presented a similar project to the Académie of Rouen in 1744. He promised nothing less "than a demonstration by machines of the principal functions of the human body," even to the point of having human maladies (Doyon & Liaigre 1957:300). According to a transcript of his first presentation, Le Cat "brought his project before the senses of the Academy by showing to it several plates" in which were depicted, among other things, the "springs" used in the machines.

[41] Aristotle holds that the two heats differ (Aristotle *Gen. anim.* 2c3, 737a1; see Lloyd 1992:153). Among Aristotelians, the question was much disputed. Suárez and Toletus both argue that it does not. Among medical authors, Fernel holds that it does (Fernel *Physiol.* 4c1); see Mendelsohn 1964.

Heat in bodies, as the first chapter of *Le Monde* tells us, does not resemble the feeling of heat or any other mode of thinking substance. Nothing in the animal-machine resembles the soul. Its resemblance is to our body alone. The principle of animal life is, as for some of Descartes' predecessors, heat; but heat is simply the violent, chaotic motion of particles. Self-motion, in the guise of self-perpetuating cycles, can be accounted for without appealing to faculties or active powers. The accomplishment of the *Traité de L'Homme,* whatever its errors, was to construct a comprehensive argument for that claim, mapping the phenomena of life onto the behavior of "mechanical things." [42]

[42] Descartes' physiology illustrates nicely the dictum of Feyerabend that every theory is born refuted. The question is whether the theory looks promising enough to fix, and when the fixes begin to look unseemly and *ad hoc.* The example of cold-blooded fish is a case in point. Descartes revises his theory of the heart in two ways: first by emphasizing the role of the blood alongside that of heat, and second by an experiment purporting to show that the blood of fish differs in kind from that of warm-blooded animals. In fact the results he reports seem to me rather dubious, but no more so than similar reports by others at the time.

[2]

Where Do Machines Come From?

Descartes was not known for his diffidence. But one problem stymied him time and again, from the years of *L'Homme* to the end of his life. In 1632 he tells Mersenne that "for a month I have deliberated over whether I would describe how the generation of animals occurs in my World, and finally I resolved to do nothing about it, because it would detain me too long" (To Mersenne Jun 1632, AT 1:254). The *Discours* acknowledges the setback: at the time he was working on *L'Homme,* he didn't know enough "to speak of [human and animal bodies] in the same style as the rest, namely, by demonstrating effects from causes, and showing from what sort of seeds, and in what manner, Nature must produce them." He therefore contented himself with a man-machine already made by God. In 1639, on the other hand, in one of his periodic bouts of enthusiasm, he writes, "I have found nothing [in anatomy] whose formation by Natural causes I think I cannot explain in particular, just as I explained, in my Meteors, the formation of a grain of salt, or of a little star-shaped bit of snow"—explained, that is, according to the "exact laws of Mechanics," imposed by God (To Mersenne 20 Feb 1639, AT 2:525). Yet he published nothing. The *Principles,* published in 1644, do not even reach the point of discussing plants and animals.[1]

---

[1] Baillet writes that during 1645 Descartes devoted "all his expenditures and faculties" to "anatomical operations," having animals of all sorts brought to him for dissection. With the aim of writing an "accurate treatise on animals," he wrote memoirs on his observations, which were communicated to Regius and used by him in what Baillet calls a "very imperfect draft" to be added to his *Fundamenta* (Baillet *Vie* 2:272–273; see also AT 4:247). In a letter to Chanut in 1646, Descartes writes that he is growing plants in his garden for experiments, in order to "continue my physics" (To Chanut 15 Jun 1646, AT 4:442).

Sometime after 1647, he returned to the problem. A letter to Élisabeth early in 1648 shows him working on the second part of the *Description,* entitled *De la formation de l'Animal.*

> I have now another piece of writing in hand, which may, I hope, be more agree-able to Your Highness: a description of the functions in animals and men. [. . .] I have even ventured (but only in the last eight or ten days) to explain the fash-ion in which the animal is formed from the start of its origination. I say the ani-mal in general; for man in particular, I do not dare try, for lack of *expérience* for this purpose. (To Élisabeth 25 Jan 1648, AT 5:112)

In his conversation with Burman three months later, he mentions an *Animalis Tractatus* (AT 5:170); late that year, he writes that "in meditating on [the for-mation of animals], I have discovered so many new lands that I have almost no doubt that I could complete the whole of Physics according to my wish, provided I had the leisure and the wherewithal to perform some experiments" (To ****, 1648 or 1649, AT 5:261). But time and means were not forthcom-ing. The *Description* too is unfinished. It omits topics that in the *Primæ cogita-tiones,* a collection of notes mostly written before 1648, Descartes had en-deavored to explain—notably the differentiation of the sexes.

The common theme in Descartes' reports is that he has not had enough time or opportunity to demonstrate the formation of animals from "Natural causes." In the *Description* he writes, "I have not until now undertaken to write my sentiments touching this matter, because I have not done enough experiments to verify all the thoughts I have had about [generation]." Never-theless he believes that the "general things" he proposes will not be belied by new experiments (*Descrip.* §27, AT 11:252–253). Less than two years later, he was dead, and the *Description* was published posthumously in 1664. One may doubt whether Descartes would ever have published a new *Traité de l'Homme,* experiments or no. The difficulties he faced were not merely practical, but conceptual.

Those difficulties were only too evident to his successors. Hermann Boer-haave said that he could not read *De la formation de l'animal* without laughing; in his physiological works "Descartes is no longer Descartes."[2] But Boer-haave's own method is resoundingly Newtonian, if not Cartesian; he too ad-mitted ignorance in understanding "the power that agglomerates the elements dispersed in the structure of the seed,"[3] and was eventually seduced into a play

---

[2] See Roger 1971:106n265 and Daremberg 1870:705n1, quoting Schultens *In mem. Boerhavii* 35.

[3] Duchesneau 1982:110.

of analogies not unlike that so often deplored by the critics of "l'homme de Descartes."[4] The nonmechanistic entities introduced by seventeenth- and eighteenth-century physiologists—plastic natures, monads, vital forces—have all gone the way of the animal spirits and the cardiac furnace. They have been replaced by material powers unsuspected by Descartes, and no doubt inadmissible in his natural philosophy. Evolution and chemistry have contributed modes of reasoning of which he had no inkling. Nevertheless, the program now pursued bears, granting all those differences, the imprint of Descartes': it too rejects teleology and nonlocal causes, it embraces capacities conforming to law, and—not least significant in the economy of science—it too offers the promise and threat of control.[5]

That is not to say that the ridicule of Descartes' successors had no basis. There is something frightening in the assurance with which Descartes writes that "those who know what I have explained about the nature of Light [. . .] and the nature of colors [. . .], will be able easily to understand why the blood of every animal is red," and then proceeds to construct, on the basis of a false theory of color, and a false theory of the elements, a likewise false theory of the redness of blood (*Descrip.* §31, AT11:255–256); or in the tranquil classification of the particles that course through the body into aereous particles and spirits, and of spirits into the unusually small and the unusually large, the smooth and the irregular . . . No doubt those entities can, if one doesn't probe too deeply, be known distinctly; they were acceptable *candidates* for explanation (*Descrip.* §3, AT11:227). But still, as Descartes himself acknowledges, God could have brought about the phenomena we experience in a near infinity of ways, which only experiments may distinguish if they can be distinguished at all.[6] One might think that, as hypothesis is piled on hypothesis,

---

[4] Duchesneau 1982:113–114.

[5] Descartes opens the *Description* with the promise that "many precepts of which we may be quite assured, both for curing and for preventing disease, and even to retard the course of aging, could be found, if people had striven to know the nature of our body, and if they had not attributed to the soul functions that depend only on the body and the disposition of its organs"; at the close of the introduction he writes that "knowing distinctly what in each of our actions depends only on the body, and what depends only on the soul, we should be able to be served better both by the one and by the other, and to cure or prevent their diseases" (*Descrip.* §3, AT 11:224, 227).

[6] The power of Nature is "so ample and vast," and the principles of Descartes' physics are "so simple and general," that he can find almost no effect of which he does not know "that it can be deduced in several diverse ways." That diversity presents the greatest obstacle to the knowledge of nature (*Discours* pt6, AT6:64–65). The same occasion for doubt is mentioned at the end of the *Principles*, which nevertheless claim "moral certainty" for the hypotheses they set forth (*PP* 4§204–205, AT 9/1:327–328).

there would be grounds for hesitation: the body is vastly more complicated than a ray of light, or even the *tourbillons* that drive the planets. Yet no such decline of assurance can be seen. In explaining the formation of animals in general, he sets down only what he expects will endure after further experiments; but of that he gives every appearance of being confident. Thus was engendered a crowd of hostages to fortune.

2.1 The First Circle

We begin with the seed.[7] The "figure and arrangement" of its particles Descartes does not describe. There is but one distinction among seeds. The seed of plants is "hard and solid." It does not change shape without becoming useless. The seed of animals, on the other hand, is "quite fluid." "Produced ordinarily by the conjunction of the two sexes," it "seems to be only a confused *mélange* of two liquors, which serve as leaven to each other," heating themselves up into something like a *feu sans lumière*.[8] Soon, "acting in the same way as new wines when they boil, or hay stored before it is dry, [the heat of the seed] causes some of its particles to gather toward some part of the space that contains them; and, expanding there, they press on the others that surround them, which begins to form the heart" (*Descrip.* §28, AT 11:254).[9]

Then, some of the expanding particles, tending, by the first law of motion, "to continue their motion in a straight line," and being resisted by the nascent heart, "move away a bit, and make their way toward the place where afterward the base of the brain will form." In doing so, they displace others, which, fol-

---

[7] I follow the account in the *Description*; my thinking has been greatly aided by Pichot's interpretation (Pichot 1993:377–388). The *Primæ cogitationes*, which are undoubtedly earlier, differ on a number of points. See Roger 1971:146–150 for a summary. In English, one of the few detailed treatments of generation in Descartes is Carter 1983, an ambitious work marred by historical misunderstandings and tenuously grounded interpretations (see also Carter 1985). Carter's work has the great merit of taking the theory seriously, and of attempting to understand it in its own terms. Unfortunately much of the discussion of generation is devoted to a detailed analogy between the formation of the solar system and that of the human body, an analogy that rests on the most slender of bases (see note 20 below).

[8] If there is seed from only one parent, it will "easily slip back by the same way that it came in: since there is nothing to keep it" in the womb (*Primæ cog.* AT 11:507).

[9] See also *Descrip.* §72, AT 11:281. There may be a reminiscence here of the Hippocratic doctrine that because of its heat, and the resulting vapor, "the seed, inflated, surrounds itself with a membrane; around it extends the exterior part, which is continuous because of its viscosity. It is thus that on baked bread there extends a thin membranous surface" (Hippocrates *De nat. pueri*; see Pichot 1993:21). The surface formed, however, is that of the body, not the heart.

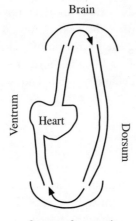

Brain

Ventrum

Heart

Dorsum

Organs of generation

Figure 3: The first paths in generation: the *Ur*-Animal. Compare AT 11, Fig. 1 (see below §4.2, Fig. 12 (i)).

lowing the general rule of motion in Descartes' world, take over the place where the first particles were, near the heart.[10] In its proximity they are heated, and "take the same path" as those that preceded them; and so a circular motion begins (*Descrip.* §28, AT 11:254). The blood begins to form from particles which are compressed near the heart, which, "crimping and dividing into very small branches," remain so near one another that only the tiny first-element particles can fit between them.

The new blood, heated and expanding as in the adult animal, moves in a straight line "toward the place where it may go most freely," and so forms the brain. The path it takes begins to form the upper part of the aorta or *grand artère*. But sooner or later it meets resistance, not from the sides of the womb as one might think, but from other particles of the seed, and is deflected back toward the heart.[11] It cannot take the path it came by, because more blood is moving along that path. So it "turns away toward the side opposite to the side by which new matter enters the heart," which will eventually be the dorsum of the new animal. Traveling downward, it forms the inferior part of the *grand artère,* and is deflected once more, now to travel toward the heart along what will be the ventrum of the new animal, thereby forming the *vena cava* (*Descrip.* §28, AT 11:257).[12]

---

[10] "All places are filled with bodies, and the same parts of matter always occupy equal places; it follows that no body moves except through a circle, so that it expels some body from the place it enters, and this body some other again, and so on to the last, which enters the place left behind by the first at the same moment that the first body leaves it" (*PP* 2§33, AT 8/1:58).

[11] The womb has only a superficial role. Quite literally: it occasions the formation of the skin (*Descrip.* §74, AT 11:284).

[12] The *Cogitationes* tell a different story. The seed is at the outset separated into subtler and thicker portions. The subtler, which move more quickly, occupy the most remote part of the womb, while the thicker tend to stay near its entrance. Eventually the subtler particles congregate to form the brain, leaving in their train the medulla of the spine, while the thicker begin to form the skin, the abdominal organs, and the lower limbs (*Primæ cog.* AT 11:508). The process is one of fractionation, rather than of circular journeys by heated particles.

2.2 The Nervous System

The reader may begin to sympathize with Boerhaave. But there is more: the formation of the nervous system. Descartes has distinguished "two sorts of particles in the portion of the seed that expands in the heart [. . .]: those which move apart and separate freely, and those which join and attach themselves to each other" (*Descrip.* §28, AT 11:258). After applying that distinction to explain why some animals have but one chamber in their heart, and others two, Descartes says that some of the particles that expand freely are *aereous,* because the bodies where they are "remain rarefied, and cannot easily be condensed" (260). Those particles, not surprisingly, form the lungs (259).

There is another kind of particle, "more lively and subtle, like those of brandy, acids, or volatile salts," which cause the blood to dilate but "do not prevent it from condensing promptly afterward" (*Descrip.* §28, AT 11:260). Such particles, "quite solid and quite agitated," are the *spirits.* Unlike aereous particles, they do not tarry for long in the lungs, but go further, into the aorta, and toward the brain. Like the blood which rises toward the brain, they are eventually deflected, "and turn to the right and left toward the base of the brain, and toward the front, where they begin to form the organs of sense" (261). Some of the aereous particles make their way along the same route.

The primary difference between blood-particles, and the aereous particles and spirits, is size. Some exceed the blood-particles in largeness, others in smallness. Of those that exceed them in smallness, some of the aereous particles have "very irregular figures that impede them," while others have "more unified and smoother figures, so that they are more suited to compose waters than air" (*Descrip.* §28, AT 11:262). The irregular ones take "the lowest path of all," and begin to form the organs of smell; the smooth ones move straight toward the front of the head, and begin to form the eyes.[13]

Of the particles that exceed those of the blood in largeness, some too have irregular, unequal figures, and are carried along by the subtle matter; and, "having more force than all the rest, because they are more massive, they move away from the center of the brain by the shortest path, and betake themselves toward the ears, where, taking along some aereous particles, they

---

[13] It is curious that Descartes here permits himself to use the word 'low' (*bas*). In humans the fetus is normally head downward in the womb. Words like 'low' and 'descend' can refer only to the location of the heart: as particles lose speed, they are entrained in circular motions leading back to the heart. Descartes needs to have the body develop asymmetrically both back to front and top to bottom; any asymmetry must be accounted for by the "fractionation" of the initial flow. Needless to say, that is a tall order to fill.

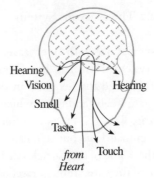

	Smallness	Largeness
Irregular	Smell	Hearing
Smooth	Sight	Taste

Figure 4: Formation of sensory nerves.

begin to form the organs of *hearing*." Likewise some have smooth figures, which make them easy to move in concert, and "consequently, with a slower motion that the rest of the spirits; and that makes them descend by the base of the brain toward the tongue, throat, and palate, where they prepare the way for the nerves that will be the organs of *taste*" (*Descrip.* §28, AT 11:263).

Still others "find pores in the seed through which they can pass." Eventually they flow into all the interior parts of the seed, and "trace there the passages of the nerves that serve the sense of *touch*" (*Descrip.* §28, AT 11:264). The fourfold classification of particles, together with the inevitable *et alia,* gives rise to the fivefold classification of the external senses. No doubt the distinctions are clear enough, but their introduction, however ingenious, rests on nothing more than the need for an explanation. I will return to that point shortly.

2.3 Flows and Resistances, Fluids and Solids

Although Descartes refers to the "machine of our body" in the introduction to the *Description,* the formation of the fetus is an exercise in fluid mechanics more than in the kinematics of rigid bodies.[14] In *L'Homme* already the man-

---

[14] "The embryogenetic model, with its streams and fluxes, is very close to the model of development and growth [. . . .] This whole aspect of Cartesian biology has in general been some-

machine is compared, not to a clock, but to a statue moved by flows of water, an organ whose pipes are filled with air. In the *Description,* Descartes holds
that "all the living parts of every body which is sustained by nourishment—
namely, animals and plants—continually change; so that there is no difference
between those that are called *fluid,* and those that are called *solid* [. . .], except that the particles of [solid parts] move much more slowly than those of
the [liquid parts]" (*Descrip.* §28, AT 11:247). The solid parts are composed of
fibers, produced "from some location on one of the branches of an artery."
Along those fibers humors and spirits are conducted, so that the body is perforated by "an infinity of little streams" originating from arterial pores. The
ends of the fibers are continually being broken off, even as sticky parts of the
blood are added to them. *Every* part of the body, at least until old age begins
to petrify it, undergoes a constant exchange of matter with the blood. This
passage deserves to be quoted in full:

> the humors and spirits that run along the little fibers that compose the solid parts
> [of the body], make the fibers advance continually little by little, though quite
> slowly; so that each of their parts has its course, from the place where they have
> their roots to the surface of the members where they terminate; and having
> reached [that surface], its encounter with the air, or bodies that touch the sur-
> face, separates the fiber; and as the extremity of each fiber detaches itself, another
> attaches itself to the root. [. . .] The part that is detached evaporates in the air,
> if it comes out of the outside skin; and if it comes from the surface of a muscle
> or some other interior part, it mingles with the fluid parts, and goes where they
> go, and sometimes [goes] through the veins toward the heart, where (as it often
> happens) they return.
>
> Thus one may see that all the parts of the little fibers that compose the solid
> members of the body have a movement which does not differ at all from that of
> the humors and spirits, except that it is much slower, just as the movement of
> the humors and spirits is slower than that of more subtle matters. (*Descrip.* AT
> 11:248)

Descartes thinks the point worth stating twice: fluidity is only a matter of degree. To think only of clocks and wind-up toys in imagining a Cartesian automaton is to be quite misled: the "machine of the body" is much more like
a petroleum distillery, or the system of pneumatic tubes for delivering mes-

---

what neglected, because it does not correspond with the idea of the body as a mechanical automaton" (Pichot 1993:381). On fluid mechanics, see Gaukroger 1995: 237, 247, 277.

sages that one still sees in some old buildings—but one which is constantly sloughing off and snapping up slivers of glass and metal.[15]

The first part of Descartes' embryology, correspondingly, describes a fluid body, a body formed not from gears and levers, but from flows and resistances. Only later do we learn about the formation of solid parts. "When the arteries and veins begin to form, they do not yet have skins, and are nothing other than little streams of blood that extend here and there into the seed" (*Descrip.* AT 11:274). Of the fluids classified in the Fourth Part of the *Description,* only certain particles of blood (those that give it its color) "properly serve to compose and nourish the solid parts [of the body]." (274). The blood, as a viscous liquid, consists in branched particles that tend to hook together. But some lose their branches as they pass and pass again through the heart, and are no longer dragged along with the majority. Yet still they retain a few branches, and so they "come to a halt against the surface of the conduits through which [the blood] passes, and thus they begin to compose the skins [of those conduits]." (275). Until now the conduits have been mere paths through the less resistant parts of the seed, like rivulets in water-soaked sand. As they take shape, other pruned particles, pushed out of the vessels by the flux of "serosities" and "vapors," form the fibers of the solid parts of the body, so that eventually we have the finished machines of *L'Homme.*

Local processes of formation before birth, in fact, differ little from processes of growth after birth.[16] The main difference is that *in utero* the most sluggish and solid parts of the seed with which development begins are nevertheless quick and fluid by comparison with the structures that exist at the end of development. Life begins in entire fluidity, and if one were to live forever, might end in entire solidity.

2.4 Descartes' Tasks

It is easy to lose patience with a story more fabulous than the acknowledged fable of *Le Monde.* The narrative flow of this tale of wandering spirits sweeps

---

[15] Descartes' next paragraph thoroughly establishes his postmodernist credentials, for in it we find not only flux but speed: "And the different speeds [with which the parts of the body flow] are the reason why [*sont cause que*] the diverse solid and fluid parts, rubbing against each other, diminish or grow, and are arranged differently, according to the different temperament of each body. So that, for example, when one is young, because the little fibers that compose the solid parts are not yet very tightly joined to one another, and the streams through the fluid parts flow are rather large, the movement of the little fibers is less slow than when one is old." (*Descrip.* AT 11:248–249).

[16] See Pichot 1993:381.

the reader into agreement. Slow it down a little, and doubt creeps in from all sides. I will not waste time on objections and replies. What is indefensible deserves no defense. Instead, consider the task Descartes has set himself, enormous, perhaps unprecedented.

1. Organization Without an Organizer

A particle of blood is merely a bounded region of extended stuff. The only "information" it carries is its *present* quantity of motion and the determination or direction of that motion. Those two properties entirely determine the outcome of any collision the particle enters into; collisions, moreover, are the only interactions among particles. Particles therefore exhibit neither memory nor foresight. The organization of the body—the construction of the network of arteries and veins, of the nerves, and finally of the solid parts—must be explicable in entirely local terms. The here and now alone matter: no particle can see what any other is doing; none can cooperate with its fellows to build a vessel or organ. For the Aristotelian, that alone would condemn the extension of Cartesian mechanism to living things. What could be more obvious in experience than that the parts of animals are fitted to one another, and develop accordingly? The heart does not grow without lungs and liver; we do not see claws on animals whose teeth and stomach are suited to a diet of greens.

Descartes too, no less than his predecessors, treats the body as consisting of systems that work together. The heart is useless without its vessels. Since the blood is inevitably depleted, without the digestive system it would soon have no blood to cook. The digestive system in turn presupposes one by which the animal senses and ingests food. An Aristotelian can think of those systems as being governed by a single soul, and a hierarchy of ends. Their development is overseen—metaphorically—by the form of the seed, which disposes the matter in the womb so that it will serve the powers of the soul to come. The static unity of the organism is its form; the dynamic unity is the directedness of its actions to a common end, the preservation of the individual and the perpetuation of the species.

We have seen part of the answer in the explanation of the pulse and the other cycles of the automaton. But the *Traité de l'Homme* presupposes an already formed machine. The formation of the machine requires a profounder application of the same modes of explanation. Now the emerging cycles of operation must not only *reproduce* the conditions of their initiation, but *produce* them—and the structures themselves that operate. Descartes' answer is that form follows function: not because the function is an end toward which the operations of the embryo are striving, but because the local efficient-

[41]

causal interactions within the seed *prefigure* the functions of the structures they eventually produce. The initial jet from its hottest part is the first segment of the circular path which will eventually form the system of arteries and veins. That jet, caused simply by heat, is already—without knowing it!—the aortal flow. Later, the skin of the vessel, the artery itself, likewise comes about by way of purely local interactions between pruned particles of blood.

The fitness of a structure to its function is to be explained, not by reference to ends, but by a kind of harmony between the efficient causes that build it and its subsequent operation.

> As more blood is made in the heart, it dilates with greater force, and by that means [the blood] advances further. And it cannot move forward except toward the locations where there are parts of the seed disposed to cede their place, and thus to flow toward the heart via the vein joined with the artery through which the blood comes, because they can have no other path but that one. Thus are formed two new little branches, one in the vein, one in the artery, whose extremities are conjoined, and which together will occupy the place of the small parts of the seed. Or else it makes the branches already formed grow longer, without any separation of their extremities. And insofar as all the small parts of the seed are thus suited to flow toward the heart—or if there are some that are not, they are easily pushed toward the surface [of the seed]—there are no parts of the seed beneath this surface in the space where the spirits are spreading that do not, in their turn, flow toward the heart. That is the reason why the veins and arteries extend their branches in every direction, each as far as the other. (*Descrip.* §48, AT 11:266)

Arteries and veins are connected, not *in order that* the circulation should occur, but *because* the advancing flow from an artery will find its easiest path by connecting to a nearby vein.[17] The body is everywhere supplied with blood vessels not in order that all its parts should be nourished, but because eventually much of the seed is displaced. Compare this with a passage from Galen:

> Furthermore, it was right of nature to bring it about that not only arteries but veins should exist in the bodies of animals. This was done so that the liver should

---

[17] Descartes says that the vein will branch too, knowing that veins are indeed branched. But the cause of their branching is mysterious. The process is like that of steam or magma forcing apart layers of rock: the cracks form where the pressure is, they don't start from somewhere else to meet it. There seem to be two analogies at work here, the one just mentioned, and an analogy between the flow of blood back to the heart, and the flow of rainwater to the sea, where creeks gather into streams, streams into rivers, and so forth. But that would require some sort of attraction by the heart akin to gravity.

be nourished by the veins alone, and by the most porous and finest of them, while the lungs are nourished by arteries, since the veins that nourish it [i.e., the branches of the pulmonary artery] are like arteries. [. . .] The providence of Nature is here to be admired, which produces two kinds of vessels at the same time, and at the same time makes openings between the ends of each which are near those of the other, and before that in the ventricles of the heart.[18]

In Galen, the two kinds of vessel exist to provide two kinds of nourishment to the various parts of the body; the generation of the venous system is, in the embryo, correspondingly independent from that of arterial system.[19] Communication between arteries and veins, especially in the cardiac septum, allows the nourishment that venous blood takes from the stomach to reach the arteries: "As the stomach is to the veins, so are the veins to the arteries" (ibid. 496). In Descartes, the veins, which are simply a return path for the blood, are formed by that very movement of returning itself, and the formation of veins and arteries is a result of that one movement. Evidently the circulation of the blood is a good thing; otherwise the blood, since it must dissipate after leaving the heart, would have to be replaced every time the whole volume of it had left. But the existence of the circulation, like that of vortices in planetary systems, is a straightforward consequence of the general rule that all motion takes place in a closed curve.[20] The alleged purpose of the veins is superfluous in explaining their formation.

---

[18] Galen *De usu part.* 6c17, *Opera* 3:495. See also 16c2, *Opera* 4:267.

[19] Although Descartes does not distinguish arterial from venous blood (except in degree of heat), arterial blood nourishes the body "wherever there are arteries accompanying the veins," the exception being the liver (*Descrip.* AT 11:246).

[20] This does provide some basis for the analogy between cosmology and physiology argued by Carter. But as I said earlier, I do not believe there is any firm basis for the detailed version of the analogy he develops. Carter cites three texts to support the claim that Descartes knowingly pursued the analogy (Carter 1983:190–191).

  (i) A letter from 1648 in which Descartes, discussing the *Description*, also responds to criticisms of the account of lunar motion in the *Principles* (AT 5:260). "This is," Carter writes, "surely an unexpected place to find a discussion of *anatomy*" (190). But many of Descartes' letters treat unrelated topics in response to questions from his correspondent. The letters to Mersenne amply attest to that. The passage cited by Carter begins, "Pour la description de l'animal" ("As for . . ."), a standard phrase by which to pass on to another topic, not necessarily related to the one preceding (see, e.g. To Mersenne 11 Mar 1640, AT 3:43; 6 Aug 1640, AT 3:144, 146; etc.).

  (ii) A response to Burman in which Descartes defends his supposition about the initial state of the cosmos against Regius (AT 5:170–171). He is ready to swear that only *after* making the supposition in order to explain celestial phenomena did he notice that fire, magnetism, and many other things could be explained by it also. He continues: "Quin etiam in ipso animalis tractatu, in quo hoc hieme laboravit, id animadvertit" ("And furthermore in the very treatise on animals, on which he [i.e., Descartes] labored this winter, he noticed the same thing"). Descartes goes on to say that although he had undertaken only to "explain the functions of the animal," he had found it nec-

Nowhere is this clearer than in the genesis of the sensory nerves outlined earlier. Yet nowhere does the gratuitousness of the particulars of the explanation strike the reader more. The sense of smell is formed by irregular, exceedingly small, aereous particles, which, impeded by their irregularity, are slowed down and "fall" toward the heart, though not so far as the grosser particles that produce the sense of taste. In *L'Homme* we learn that

> when the machine breathes, the more subtle parts of the air that enter it through the nose, penetrate through the pores of the bone which is called *spongiosum*, if not as far as to the concavities of the brain, at least as far as to the space which is between the two membranes [the *pia mater* and the *dura mater*] that envelop it [. . .]; and at the entry of this space they encounter the extremities of the little threads [in the olfactory nerves], which are either bare or covered with an extremely fine membrane. (*L'Homme*, AT 11 : 148)

The pores of the *os spongiosum* are "so disposed, and so narrow" that they allow no particles of the third element to pass to the olfactory nerve-endings except those that Descartes calls "odors." Unfortunately we are not told here what sort of particles odors are, except that only particles somehow out of the

---

essary to explain its formation from the egg also, "which he noticed [advertit] followed so well from his principles that he could give a reason why the eye, nose, brain, etc. existed," and indeed could not exist otherwise than they do, "from his principles." The parallel is: just as I found I could explain more than I'd expected from my supposition about the initial state of the cosmos, so too I found I could explain not only the functions of the animal but its formation. Now perhaps Descartes did believe that in pursuing the consequences of his principles, he could construct a continuous story leading from the initial state of the world to the actual configurations of animal bodies. But as it stands the response indicates only that he thought that the formation of animals could be explained from his principles. Whether that would be in analogy with the formation of the heavens is unclear. 'Quin etiam' means 'moreover'; it need not imply any consequence-relation.

(iii) At the beginning of the Fourth Part of the *Description*, Descartes writes that he hasn't performed enough experiments to verify "all thoughts I have had" on generation; but he will put something down which is general enough that "je serai le moins en hazard cy-après de me dédire, lors que de nouuelles experiences me donneront davantage de lumière" (AT 11 : 191). Carter translates the last phrase "when certain experiments give me further information concerning light." But 'davantage de lumière' means 'more light,' i.e., more *insight*. The passage parallels several others mentioned earlier in which Descartes indicates he must do more experiments in order to settle his views on generation.

Neither this nor (i) offers any support for the analogy developed by Carter, and (ii) is equivocal at best. It would be surprising if Descartes had not somewhere explicitly indicated that there was such a connection—he could well have done so in the very passage from Burman quoted above, or in the *Description*. But he doesn't.

ordinary will excite the sense of smell. It is reasonable to identify them with the irregular, small, aereous particles whose path in the embryo will become the olfactory nerves.[21]

The sense of smell is made *by its medium.* So too is taste. The auditory nerves are not fashioned by sound, but the particles that make them are grosser versions of the particles that produce odor, that is, they are particles of what we ordinarily call *air,* which is not sound but the medium of sound. Similarly the sense of vision is produced not by light, nor even by the second-element particles which are its medium, but by the spirits that most resemble the second element—exceedingly small, smooth, aereous particles.

The fact that a sense is manufactured by its medium or something like its medium has little to do with its operation in the fully formed organism. The nervous system is the same everywhere, and the sense organ itself consists in a more or less exposed set of nerve-fibers. It is as if Descartes, needing *some* distinction in their causes, could think of nothing better than to invoke a pe-culiar version of the Aristotelian theory that each sense resembles the medium of its *sensata.* The unswerving aim of his physiology is to show how the body is made—the structure *and* the process—without ever mentioning what it is *for.* Even the weakest hypothesis about mechanical causes is preferable to the ascription of ends.

2. Transmission of Characters

The seed begins as an undifferentiated mass, heated by fermentation. Certainly there seems to be no means by which it could embody information about the specific or individual characters of the parents. Descartes does write that

> insofar as the little threads of which the solid parts are composed are turned aside, fold up, and intertwine in various ways, according to the various flows of fluid and subtle matter that surround them, and following the figure of the places they encounter: if one knew well all the parts of the seed of a particular species of an-imal—*man,* for example—one could deduce from that alone, by reasons en-tirely mathematical and certain, the whole figure and conformation of each of its members; as conversely, knowing some of the particularities of the confor-mation, one may deduce that of the seed. (Descrip. §48, AT 11:277)

---

[21] Smell and taste were traditionally regarded as very similar in their objects, a view taken over, for example, by the Cartesian Johannes Clauberg. "*Odors* are generated and distinguished in almost the same ways, and are called by the same names, as tastes. [. . .] But there is this dif-ference between them, that odor consists principally in dry tenuous parts, taste in grosser and more humid" (Clauberg, *Theoria corpora viventium* §417–418; see also §900; *Opera* 1:180, 200).

The seed, in other words, must (despite its being said to be "quite fluid") contain regions of higher or lower density; or perhaps, since in the Cartesian plenum all bodies are equally dense, variations in the ratios of the first, second, and third elements; or perhaps variations in viscosity constitute the "figure" or "conformation" of the seed. All we know is that the seed resists the flow of the proto-blood differently in different places. Little wonder that Descartes has nothing to say.[22]

In fact, although Descartes proposes that one may deduce the conformation of the seed from that of the animal it produces, he earlier gives a methodological reason for not doing so, at least not immediately. "in order that the knowledge one has of the figures of animals already formed should not hinder one from conceiving [the figure] they have at the beginning of their formation, one must consider the seed as a mass, from which is first formed the heart." (Descrip. §48, AT 11:264). A general schema, consisting of the heart, the *vena cava* and the aorta, the future locations of the head and the lower body, and finally the spine which connects the two, will be derived merely from the fermentative heat of the seed (see Fig. 3). The schema is a kind of Ur-animal, a ground plan that follows merely from the most general properties of the seed.[23] Only after having demonstrated that schema will the embryologist venture to derive the characters of particular species, and attempt some theory of their transmission. Descartes does not solve the problem of the transmission of characters. Far from it. But at least he divides the question, separating what will take place in every gestation from what will occur only in some.

Perhaps the closest parallel in Descartes own work is the theory of the formation of the Earth in the *Principles*.[24] Though he doesn't say so, it is in fact

---

[22] Although Descartes' embryology is certainly epigenetic by comparison with Malebranche, Perrault, and Régis (see Roger 1971:334–353, Pichot 1993:386), it can hardly avoid all suggestion of a pre-existing formative configuration. Descartes' quondam disciple Regius was not wholly unorthodox in holding that the seed includes "the engendered rudiment of a similar animal [i.e., to the parent]," a figure in the particles of the seed that "forms a germ or sketch" of the animal to be engendered (Roger 1971:153, quoting Regius *Fund. phys.* 208, 209). But in Descartes' scheme the seed is more like a mold than a preformed individual that has only to grow to adult size; indeed it would seem to contain primarily the form of the circulatory system, the nerves, and perhaps the outline of the Ur-Animal in Fig. 3.

[23] Thus when Descartes writes near the beginning of *Description* 4 that he is going to "put here in passing something of what is most general," and in other passages where he refers to the "animal in general," the Ur-Animal is that animal. It is logically quite unlike the *genus* 'animal' or a Platonic idea of animalness. Whether such an entity had ever been conceived before in natural philosophy I do not know; afterward, of course, there were others like it—Goethe's *Urpflanz*, for example, or Cuvier's prototype.

[24] See *PP* 3§94, 119, 146; 4§2–14, 32–44; AT 8/1:147, 168, 195, 203–208, 218–231. 4§2 is a summary of the transformation of stars into planets, with the Earth as instance.

a theory of the formation and early history of any planet. Nothing distinguishes the Earth except that like some but not all planets it has a planet of its own, the Moon. But that has no bearing on its development. What Descartes is describing is the formation of "the planet in general," the *Ur*-Planet. We are not surprised to find that ends are dispensed with in the theory of planet-formation. The three layers of the early Earth arise straightforwardly from local efficient causes (*PP* 4§3; 204–205), as do further layers (§34, 37–38; 220–221, 224–226), mountains, plains, oceans, and so forth (§44; 230–231). There is no reason why every planet of sufficient size and solidity should not have the same structure. Similarly, there is no reason why every animal should not have the same basic body plan.

3. Sexual Difference

The *Description* explains the formation of structures found in all humans. It mentions aging, and leanness and fatness. But those are postnatal changes. Only in the Latin notes published posthumously as the *Prima cogitationes circa generationem animalium* do we find an explanation of the most striking of individual differences: the difference of the sexes.[25] In fact the *Cogitationes* offer two incompatible explanations.

Any account of individual difference must introduce a variable factor, a wild card, absent from the accounts of features found universally. In the first explanation, that factor is geometrical.

> The fetus, on account of the sympathy of its movement with the mother, sends forth a penis as from the back of the mother: that is, with its root toward the back of the mother, and terminating toward her navel. Hence if the head of the embryo is toward the navel of the woman, while its buttocks are toward the spine of the back, the fetus will become a male, and the penis will go outside [its body]. If, on the other hand, the head of the embryo is toward the spine, and the buttocks toward the abdomen, it will be female, since then the penis curves back toward the navel of the mother toward the interior parts of the embryo.[26]

[25] Though Descartes is unlikely to have devised them both at the same time, I see no firm grounds on which to decide which was first, or for dating either of them, though it is possible that the second (in the order of the text) was devised after Descartes realized that fetuses are not randomly oriented in the womb. In what follows, 'first' and 'second' refer only to the order in which they were printed in the *Opera* of 1701. On this text and its authenticity, see AT 11:501–504.

[26] "Fœtus, propter sympathiam motûs cum matre, emittit penem tanquam ex dorso matris: id est, radice ejus existente versùs matris dorsum, terminatur versùs ejusdem umbilicum. Hinc sit, ut si caput embryonis sit versùs umbilicum fœminæ, fiat masculus, & penis foras exeat. Si contrà caput embryonis sit versùs spinam, & nates versùs abdomen, sit fœmina; recurvatur enim penis versùs umbilicum matris ad interiores partes embryonis" (AT 11:515–516).

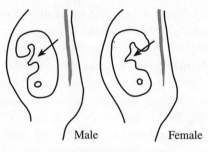

Figure 5: Sexual differentiation.

Common to male and female is the sending forth of a sexual organ, always in the direction from the *mother's* back to her front (hence, I suppose, the invocation of "sympathy"). The fetus either faces forward or backward. If it faces forward, the organ will extend outward from the fetus; if backward, it will extend inward (and in both cases upward, toward the fetus's head, since the fetus is always upside-down with respect to the mother). Hence the difference in the sex organs of the mature organism.

That the vagina is an inverted penis (and vice versa) is not a new idea. Only the explanation is new. It is so far impressively neutral. Male and female differ only with respect to orientation. But the consequences drawn from that difference are not neutral.

> From this one may conjecture why males should be more ingenious [than females]: because a purer part of the seed might be carried higher, and so be possessed of more powers. Likewise, why they should be more robust: because the spine of the fetus is nourished near the spine of the woman [i.e., the mother]. Or again, why women should have more ample posterior parts: because, being next to the abdomen of the mother, which is softer than the spine, they can more easily expand.[27]

The paragraph illustrates the comprehension of a diversity of phenomena under a single hypothesis about the cause—a procedure found elsewhere in Descartes' work. The phenomena are, some of them, dubious by our lights; and the confirmation received by the hypothesis correspondingly diminished. But Descartes is simply following his usual strategy of being conservative with

[27] "Hinc conjicere licet, cur mares sint magis ingeniosi: quia etiam pars seminis purior altiùs ferri potuit, ac proinde plus habebat virium. Item, cur sint robustiores: quia fœtûs spina alitur prope spinam mulieris. Item, cur fœminæ habeant posteriores partes ampliores: tum quia juxta abdomen matris, quod mollius est quàm spina, faciliùs possint extendi" (AT 11:516).

respect to the phenomena. Not for him the task of overturning what were then widespread, if not universal, views about men and women. His aim is only to explain them, even if—as with the *marques d'envie* that Mersenne inquires about, or the order of battle supposedly observed by apes—he himself sets little store by them. The explanation itself, it should be noted, does *not* presume any difference beforehand in the nature of what will become male and female bodies. Orientation is an extrinsic relation to the mother's body.

The second explanation follows two paragraphs on the ingestion and excretion of aliment by the fetus. Against certain Galenists Descartes argues that the fetus takes in food through the mouth and excretes waste through the same passages—once they exist—that it will use after birth. The excretions include solids, liquids, and gases (*flatus*). As before, a variable feature is the basis of sexual difference.

> It is to be noted that these feces [i.e., the first to be excreted] are gas and urine, not solid wastes (gas is as robustly produced as, or more than, other feces). Now if the fetus is stronger by virtue of a more robust nature, more urine is purged from it, than gross solids [. . .]; hence the penis is perforated first and sticks out, and the fetus is male. But if the fetus expels more of the solid excrement, and retains in itself the aqueous humors, it will be of a softer nature, and the anus will be perforated first, through which the solid waste in exiting occupies the groin, and impedes the protrusion of the pudenda, which instead protrude within, and [the fetus] will be female. If, finally, it is of an equal temperament, so that both are perforated in the same moment (which rarely happens), it will be a Hermaphrodite.[28]

The variable here is the more or less robust nature (hence the greater or lesser strength) of the fetus. The stronger fetus excretes liquid and retains solid, and becomes male; the weaker excretes solid and retains liquid, and becomes female; one of "balanced temperament" excretes both at once, and becomes hermaphroditic. Since robustness is a matter of degree, sex is too: it would be

---

[28] "Notandum: fæces istæ sunt flatus & urina, non stercus, (flatus autem efficit æquè vel magis robustum quàm aliæ fæces). Iam verò si valentior sit fœtus ex naturâ robustiori, plus urinæ ex eo purgatum est, quàm stercoris crassi (tunc sola glans emersit ex corpore, quæ deinde præputio tecta est flaccescente cute); ideoque penis prior perforatur & prominet, fœtusque est masculus. Si verò fœtus excernat plus solidi excrementi, retineatque intra se aquosos humores, sit naturæ mollioris, priusquam perforatur podex, per quem exiens solidum excrementum premit inguina, impeditque ne pudenda foras promineant, sed ea intus protrudit, & sit fœmina. Si denique sit tam æqualis temperies, ut utrumque eodem momento perforetur, quod rarò contingit, sit Hermaphroditus" (AT 11 : 523–524). The passage is marked by the editors of the 1701 edition as having been deleted by Descartes.

more proper to speak of one individual being more or less male (or female) than of its being simply male or female.[29]

Unlike orientation in the womb, robustness is an intrinsic prior difference. Males and females are different even before they acquire the organs that make their difference obvious to the senses. It is not, of course, surprising that the male is the more robust (and consequently the less humid, in accordance with tradition). Nevertheless the female is not less perfect, it would seem, in her anatomy. Only the prior connection of robustness with the excretion of fluids before solids introduces an inequality—of degree only—between the sexes. Descartes, again conservative, does not offend common opinion where it concerns the phenomena, and when he has no systematic reason to do so.

The first explanation takes the orientation of the fetus for its variable factor, a clear enough notion (but it also invokes "sympathy" between the fetus and the mother). The second employs robustness, a notion that demands clarification. But it avoids what may have come to seem a dubious correlation between orientation and sex, and allows for ambiguous sexual configurations, a possibility that the first explanation can hardly make sense of.[30] What is common to both is, first, that the explanation of sexual difference has nothing to do with the soul. There are no male or female *souls,* there are only *bodies* that happen to become one or the other as they develop in the womb. The second point is that sexual difference, even where it is referred to the degree of robustness of the organism, is superficial. It does, as Descartes observes, have consequences for the subsequent development of the organism. Men, for example, being drier, grow beards, although since old people of both sexes are drier (which is to say less fluid in Descartes' sense) the growth of a beard can occur in either sex.[31] Nevertheless—and whatever one may surmise

[29] The logic of sex, under this conception, is like that of concavity and convexity. An arc with a positive degree of curvature, however small, is convex; with however small a degree of negative curvature, it is concave; and if the curvature is zero, it is neither (or, if you like, both). Nevertheless the "dimension" of curvature is continuous: so too with sex.

[30] The difficulty of dating the two passages makes it perilous to draw any conclusions about changes in Descartes' thinking about sexual difference. In particular I do not know what to make of the change, if it is one, from an all-or-nothing conception of male and female to one in which sex is a matter of degree.

[31] AT 11: 525–526. The correlation of facial hair with dryness is not new: but the mechanistic explanation of the correlation is. The dryness of males results from the transpiration of spirits through the testes ("in illis [i.e., the testes] copia spirituum assiduè fiat, quæ per scrotum in auras transpirat"); *castrati*, therefore, as well as women, will have less facial hair than intact males. Again Descartes is content to show that he can account for the *phenomena* as common sense presents them (including the association of the sexual organs with the brain, an association we have seen in Bauhin's illustrations of the pineal gland and its environs). The point here is not to produce new phenomena, but to explain as many acknowledged phenomena as possible.

about the cultural value attached in Descartes' time to robustness, or dryness and wetness—there is no *scientific* argument for the inferiority of women. The laws of nature, which are the true cause of the motions by which the organism is formed, are quite indifferent to the condition of the bodies they govern. In the published version of Descartes' notes, the second explanation of sexual difference is followed by a paragraph which, if it is indeed a reflection on what precedes it, testifies to this point.

> I expect that someone may, with furrowed brow, say it is ridiculous for a thing of such importance as the procreation of man to be brought about by such slight causes. But what cause could be of more weight than the eternal laws of Nature? Should [these things] be brought about by some Mind? By which [Mind]? Or immediately by God? But then why are monsters sometimes brought about? Or by that most wise Nature, which has no other origin than in the folly of human thought?[32]

In the theories of generation here being alluded to, the perfection (or the nobility) of organisms requires a corresponding perfection in their causes. The souls of higher animals require for their production more than mere matter and material qualities: they must be produced by intelligent causes. The souls of humans, the most perfect of those forms that can be joined with matter, can be produced only by God. Though Descartes is implying that the laws of Nature are no less noble, considered as causes, than the celestial intelligences invoked by his opponents, those laws are universal. Indifferently governing every change in nature, they are "causes" of the generation of *every* living thing, with no distinction between higher and lower. So too with male and female: the cause of sexual difference is simply a configuration of matter, which brings about, in accordance with the laws of nature, further changes in the fetus.

I would not go so far as to say that Descartes did not at all share the attitudes prevalent among his contemporaries, or that he was prepared, as some were, to argue for the equality of the sexes.[33] That was not his way of operat-

---

[32] AT 11:524. The invoking of "Minds" or of God in generation is found in Aristotelian theories of generation, and in particular in questions on the "eduction" of forms from matter. The souls of higher animals and especially of humans are beyond the capacity of purely material causes to produce; they are instead introduced into the developing fetus by celestial intelligences or (in humans) by God himself. There is a parallel passage in a letter to Morin concerning the cause of light (AT 3:).

[33] For a very brief consideration of Descartes' physiology and gender issues, see Clarke 1999:91–93. Although I cannot agree that Descartes "excludes from his biology any type of language that could privilege the male role," I agree that because the body "is to be understood only in terms of extension" (though not only "by describing quantifiable characteristics," un-

ing. But the reduction of nature, animate and inanimate alike, to extension and its modes, and the rejection of any order of perfection within the world of material things, though it may reflect in some sense a "masculine" distancing or alienation of the knowing subject from the objects of knowledge, also offers a potent tool by which to frustrate the projection of cultural valuations into the natural world.

∞

Descartes' errors have a complex etiology. His pride, his confidence—unjustified, as it turned out—in his capacities, in the possibility of making natural philosophy the work of a single mind, was one factor. We might hark back to the *Regulæ* and their insistence on the continuous, almost instantaneous apprehension of demonstrations which alone would allow us to have confidence in their conclusions: if that is the path to knowledge, then science can hardly be the work of a community, even if in the *Discours* Descartes admits that a single lifetime might not suffice for the completion of natural philosophy.

The apparent success of certain parts of his program—the explanation of the formation of images on the retina, or of the repeated beating of the heart—did nothing to lessen Descartes' confidence. He did not distinguish so firmly as I have between a proof of concept and a proof *simpliciter*. His reliance, moreover, on the anatomy of his predecessors, and on the Galenist framework of animal spirits, gave him a false basis on which to build the animal-machine, though one must note that the animal spirits were not conclusively rejected until the eighteenth century.

Finally, the imagery of the fluid body, and of the "subtle" as a kind of material analogue to the spiritual, would seem to have blinded Descartes to the dynamical impossibilities of his system—notably in the explanation of sensation by reference to spirits issuing from the pineal gland—; his warnings elsewhere notwithstanding, he too was seduced by the imagination.

---

less by that Clarke means that those characteristics must be basic or derived modes of extension), there are in Descartes' physiology grounds for denying any *ethical* value to sexual difference. I should note also that Clarke is ignoring contemporaneous two-seed theories of generation, which gave the female a more active role in the production of offspring. On arguments for and against the equality of women in France at the time of Descartes, see Albistur & Armogathe 1977, c.5.

[3]

The Uses of Usus

The intent of Cartesian physiology is to make *usus* or function no more than a gloss on the local efficient-causal narrative that tells us how things happen.[1] The heart is where the blood heats up, and expanding, is forced out into the arteries—*and,* if you like, its function is to be the organ where that takes place. But two obstacles stand in the way of stripping off the gloss.

First is the naming of parts of the body: its organs, humors, and tissues. In the language of common sense and of Aristotelian physiology naming relies on function. A thoroughgoing rejection of teleology would require that some other basis be found by which to identify the parts of the body. The fable of imitation in *L'Homme* is a compromise, deferring the task of doing so in favor of borrowing the names we have, and basing their use for the parts of the *homme-machine* on resemblance. The *Description,* on the other hand, divesting itself of the fiction of resemblance, contents itself with commonsense names; implicitly it offers an alternative based on the efficient-causal origins of bodily structures.

The second obstacle is that the body whose parts and processes Descartes explains is a *normal* body, a well-functioning body. Yet Descartes has no means, it would seem, to define normality, except when the body is united with a soul. When it is united, the divine institution of relations between modes of the soul and modes of the body provides a standard. The body-machine is healthy just in case its operations permit the perceptions of the soul—its

---

[1] On Galen's word χρεία, translated *usus,* see Hankinson 1989:217n32. Hankinson prefers 'function,' but the Greek word "may be rendered variously as 'function,' 'purpose,' 'end,' 'need,' 'use,' and even 'reason.' The Latin word can also mean 'need,' but its central senses are 'use,' 'usage or custom,' 'utility,' and 'function.'"

sensations and passions—to function as a guide to benefit and harm, defined as the preservation or destruction of the union. But health so defined applies only to humans: to animals it would apply only *per analogiam* with our states of health and disease. The alternatives are to take the continued self-motion of the machine, or some other overall effect, *as if* it were an end, or else to comprehend within physiology *all* the states of the machine. The first alternative is arbitrary; the second runs the risk of dissolving the boundaries of physiology, for lack of any distinctions among the indefinite variety of physical changes that may befall the machine.

3.1 Naming of Parts

Near the beginning of *L'Homme,* Descartes writes: "I suppose them [the parts of the machine] to be entirely similar to the parts of our Body that have the same names, and which you can have shown to you by some learned Anatomist—those, at least, that are big enough to be seen—if you do not know them" (AT 11:120–121). There is an echo here of the Aristotelian notion of *homonymy.* The first sentence of the *Categories* defines *homonyma*—in Latin, *nomina æquivoca*—to be those things "of which the name alone is in common, but according to their names the definition of the substance is different." [2] The word 'animal' is equivocal when used both of a man and of a depicted man, because 'man' and 'depicted man' have different definitions or *rationes.* In *De anima,* a similar example appears as Aristotle explains how the soul is the form of the body: "If the eye were an animal, its soul would be vision. For this is the substance of the eye according to its definition. The eye is the matter of vision, and if vision is lacking, it is no longer an eye except equivocally, like an eye of stone or a depicted eye" (Aristotle *De An.* 2c1, 418b19–23, in Thomas Aquinas *In de An.* (Pirotta) 2lec2). Thomas, in his commentary on the passage, cites the *Categories* definition: "And it is thus, because [those names] are equivocal, whose name alone is common and the definition of the substance different" (ibid. no. 239). [3]

Accept for a moment the pretense that *L'Homme* is a fable. The name 'heart,' applied to the organ pointed out by our learned Anatomist, and to

---

[2] "Æquivoca dicuntur quorum solùm nomen commune est, secundùm nomen verò substantiæ ratio diversa" Toletus *In Log. Prædicamenta,* 1c1 and 1q1, *Opera* 2:81–82.

[3] See also Toletus *In de An.* 2c1text8, *Opera* 3:38vb. Descartes, of course, was familiar with the notion of *æquivocatio.* See, for example, *Resp.* 5, ad2no3, AT 7:356: "But I, observing that the principle by which we are nourished differs entirely in kind from the principle by which we think, said that the name 'soul,' when used for both, is equivocal."

the bit of machinery described by Descartes, is equivocal, no less so than 'eye' when used of Mona Lisa's eye and that of the spectator who returns her gaze. The bit of machinery, however, much more than the brush-strokes of Leonardo, is "entirely similar" to the organ, as similar as God can make it. So, it would seem, we have only to look in order to designate by the appropriate names all the parts of the man-machine.

But we must look *first* at the dissected body that the Anatomist has allowed us to observe. *There* is the heart, and what is very like it in the machine in the other world, *that* is the "heart" of the machine. "The living thing is given as such, prior to the construction of the machine. In other words, to understand the machine-animal, one must grasp it as being preceded, in the logical and the chronological sense, both by God, as its efficient cause, and by a pre-existing living thing which is to be imitated, as its formal and final cause." That final cause, of course, is no sooner implied than discarded, on the way to the description of the human body. *L'Homme* sets aside "only by a feint the existence of what it is supposed to represent" (Canguilhem 1975:112–113).

That existence—the human body in this world—would seem to be, nevertheless, indispensable. How else would one identify the parts of the man-machine? Or is the heart simply the hottest part of the body? The lungs the most aereous? Is blood just the viscous stuff circulating in the veins and arteries? For Aristotelians, the question falls into at least two cases.

(i) The organs of the body, by and large, cannot be identified as such except by reference to the powers they subserve. Those, even though they are not full-fledged substantial forms, play the role of forms in the definitions of organs. The *ratio* of the eye is seeing; without the soul there is no seeing; so nothing that is not part of a living thing is an eye except equivocally. Even a fresh-killed corpse has no eyes in the strict sense: the *forma cadaveris,* if there is one, does not have the power to see.

(ii) Flesh and bone, on the other hand, may well be mixtures whose *rationes* consist in a certain temperament of primary qualities, identifiable quite apart from their role in the life of the organism.[4] But they have no vital powers, no organizing function, of their own. Only the formative pow-

---

[4] Jennifer Whiting distinguishes "compositional flesh" from "functional flesh." Compositional flesh is a temperament of matter which may survive the departure of the soul. Functional flesh, which is constituted by but not identical to compositional flesh, is flesh considered in its relation to the vital operations of an organism; only a thing with a soul has functional flesh (Whiting 1992:80–81, citing Aristotle *De Gen. et corr.* 321b19–22 = Coimbra *In de Gen.* 1c5tex35). At the passage cited the Coimbrans note that "on what Aristotle here intended to signify by the terms 'form' and 'matter,' or 'formal and material part,' there is no small controversy"; they take up the question in 1c5q12, *In de Gen.* 267–271.

ers of the seed, or later those of the soul, can make organs out of them. A certain arrangement or disposition of humors and tunics is necessary to the exercise of the power of vision in humans, and thus to the existence of eyes; but an eye is an eye just in case it has the power to see.[5]

Descartes' answer, on the other hand, *ought* to be that when we compare the heart of the cadaver before us with the parts of the machine, the ground of the similarity on which we base the identification of one of those parts as the "heart" of the machine is a collection of basic and derived modes of extension—its size and shape, its characteristic motion, its hotness. Remember the eel's heart, dissected out and still beating hours later. If we saw a "heart" lying on the ground in the world of *Le Monde,* we could still rightly call it a heart, even if it had never been in a man-machine, even if it were quite cold. It is not a *working* "heart." It isn't a furnace. But to deny it is a "heart" is like saying a mainspring is no mainspring if it is not now in a functioning watch. To call only a working heart by that name would be an arbitrary decision on our part, like using the term 'bronco' only of horses that are trying to unseat their riders.

Descartes' brief remarks on spontaneous generation in the *Primæ Cogitationes* confirm the point. There are two kinds of generation, "one without seed or womb, the other from a seed."

> Every animal that originates without [being born from] a womb, requires only this principle: namely, that two subjects, not very far removed from each other, should be excited in various ways to the same force of heat, so that from one of them subtle parts (which I will call vital spirits), and from the other one grosser parts (which I call blood or vital humor) strive to issue forth; those parts meeting together produce life first in the heart, where there is a perpetual battle between the blood and the animal spirits, and then, after the blood and the spirits are so conquered by each other that they can agree in nature, they generate the brain. Since, therefore, so little is required for the making of an animal, clearly it is not surprising if we see that so many animals, so many worms, so many insects are formed spontaneously in any putrefying matter.[6]

[5] Without endorsing Miles Burnyeat's interpretation of Aristotle himself, I note its agreement with the Aristotelian position presented here. The transparency of the sensitive part of the eye and the moderate heat and hardness of the skin "are merely necessary conditions for perception to take place. They are not part of a more elaborate story which would work up in material terms to a set of sufficient conditions for the perception of colours and temperature." "The animal has the faculty of perception, the air [. . .] does not. That's all—that's where *for Aristotle* explanation comes to a stop" (Burnyeat 1992:22, 25).

[6] "Omne animal, quod sine matrice oritur, hoc tantummodo principium requirit: nempe ut duo subjecta, ab invicem non valde remota, ab eâdem vi caloris diversimodè concitentur, ita ut ex une subtiles partes (quas spiritus vitales deincepts appellabo), alio crassiores (quas sanguinem

Nothing is required to make an animal but two *subjecta* whose volatile parts have complementary physical properties, and are apt to struggle when they meet. That struggle is real enough, and constitutes, as Descartes says, "the life of the animal, not otherwise than the life of fire in a lantern" (AT 11:509); that lantern *is* the heart, lit or not.[7]

Yet there is one grave difficulty that Descartes passes over lightly, in part because his stereotypical animals are *bruta* and men. Insects and worms, the oyster—traditional archetypes of the "imperfect" animal—figure only here and there.[8] In cows and people, the heart is easy to identify; but as one moves further down the hierarchy of living things, what is called the heart becomes less and less like the human, until in some animals we may not know which organ to call the heart. Descartes himself says that in some animals the heart has but one chamber, in others two, in others three. The Aristotelian can call those organs "hearts" because the term refers to them by way of their function; a function or power, like a specific form, can be composed with different matters. In honeybees the eye is compound, and works on different principles than the eyes of cows and people. That is no bar to calling it an eye: the quasi-form of the eye is vision, not some arrangement of parts.

But can Descartes call them eyes and hearts? How like a human heart must a heart be? How near to the heart's must its relative dimensions be, its characteristic heat? Immediately after the passage on spontaneous generation cited above, Descartes writes that lungs and liver are requisite in animals, because the contending humors and spirits come from them. That must include worms and insects. So we ought to be able to designate the liver of a worm,

---

sive humorem vitalem dicam), cogat erumpere; quæ partes simul concurrentes efficiunt vitam primò in corde, ubi est sanguinis & spiritus ita fuerunt unus ab altero domiti, ut in eandem naturam possint convenire, generant cerebrum. Cùm igitur tam pauca requirantur ad animal faciendum, profectò non mirum est, si tot animalia, tot vermes, tot insecta in omni putrescente materiâ sponte formari videamus" (*Primæ cog.* AT 11:505–506). This section of the *Cogitationes* must be quite early—witness the *pugna* or "battle," later referred to as a *certamen* or "struggle" (509), in the heart between the animal spirits and the blood.

[7] The possibility of spontaneous generation—in recent literature, thought-experiments about "swampmen" and "swampcows" are structurally similar—precludes the definition of function by way of evolutionary history. No one in the period would have thought to deny that a worm spontaneously generated is a worm, though it has no parent, and no history in the required sense. Only when, through the work of Joseph Needham and Louis Pasteur, spontaneous generation was ruled out could a notion like Ruth Millikan's "proper function" begin to take shape.

[8] Swammerdam and Réaumur's work on insects, Tremblay and his polyps, the biology of yeast and planaria, are all in the future. Lamarck, looking back on the eighteenth century, writes that "What is singular is that the most important phenomena were not available to be meditated upon until the epoch when the study of the least perfect animals became the principal interest, and where researches on the different complications of the organization of those animals became the principal basis of their study" (Lamarck *Philos. zool.* 1:22–23; see also 28–30).

as the part that supplies vital humors to the heart. But that looks like a function, and a worm-liver mechanism would be a liver only equivocally.

The difficulty arises not only in comparative anatomy but in identifying the organs of one species. What distinguishes the organ of sight from that of hearing? It cannot be that one is acted on by sound alone, the other by light alone. The Aristotelian can say that, perhaps; for Descartes sound and light are just motions. Still it is only through the eyes that the motion or tendency to motion which is light comes from distant objects to alter the course of the animal spirits in the brain, and thereby change the movements of muscles. Yet here too we verge on function: what matters in an eye would not be exactly what sort of mechanism it is, but that it should link impinging images with the actions and passions of the machine. That, then, would be its *ratio,* its definition.

But not for Descartes. It is true that the Cartesian definition of an organ abstracts somewhat from matter. The "form," as Descartes calls it, of blood is a certain configuration of basic and derived modes.[9] The "fibers" whose slow but constant attrition and repair is one of the main effects of circulation are particles whose length greatly exceeds their diameter, and which overall are bigger than any of the particles in the streams around them. Organs so defined can be identified in other animals, even in worms. But when we come to the larger, more complicated parts of the body, Cartesian definitions begin to look insufficient. Not because any one eye or heart can fail to have a configuration, or because vision or forcing blood into the arteries would not follow necessarily from its having the configuration it does: but because it could have *any* configuration of which the same was true without failing to be an eye or heart. Philosophers now talk of "multiple realizability." The capacity to extract information about distant objects by intercepting the mingled waves received from them, is realized in human eyes, bee eyes, or charge-coupled devices in a video camera. Even if some baroque collection of physical predicates determined a set coextensive with the set of eyes in the world, it would still be the function that matters.

One striking sentence in Descartes' *Excerpta anatomica* hints at an answer. "It is no wonder that almost all animals generate: those that cannot generate cannot be generated, and consequently are not found in the world" (AT 11:619). *Nec mirum:* we have no cause to wonder that God has given to his creatures the power to produce their like. Whatever did not would soon be gone. There is a clear parallel with the embryology. The aorta is not formed in order to convey fresh blood from the heart to the rest of the body. Local

---

[9] *L'Homme,* AT 11:123, 130.

efficient causes produce it; if they did not, the aorta would not exist, and the animal would not live. We need not wonder that the animal has such a vessel. Were there no vessel there would be no animal. To say that the aorta is for conveying blood is only a misleading way of stating those facts. It misleads because (in Descartes' view) ends entail minds to conceive and act on them. But neither the aorta, nor its owner, nor the seed, nor the parent has conceived anything.

On the other hand, we can only *begin* with the names we have. Descartes will grant that common sense favors the Aristotelian. But starting from the commonsense notion that the eye is for seeing, the philosopher can work backwards, as Descartes undoubtedly did, to the counterfactual truth: if the machine had no eyes, it would have no vision, because the eye-mechanism is essential to making light affect the actions of the machine. Once the philosopher demonstrates that truth, little remains for the ascription of ends to do, except to provide a bridge to ordinary language or a way to begin conversations with Aristotelian opponents.

The framing of the *Traité de l'Homme* as fable undermines its purpose. In it indeed, as Canguilhem says, the living animal is prior to the machine, and serves as its model. The machine is *like* the animal, in outward form and operation. But the whole intent of Descartes' physiology is to show that the term 'heart,' applied to the animal-machine, is *not* homonymous, and that the gap is a temporary one. Its eye *is* an eye, though nothing "sees" with it in the strict sense unless the machine has a soul.[10] *L'Homme*, like *Le Monde*, is a *roman à clef*, whose whole interest stems from the identification of certain of its characters with real people. Were it to succeed as fiction, it would fail as science. But the reference to the "learned Anatomist" suggests that *L'Homme* is, even among *romans à clef*, peculiar. Its characters, it would seem, cannot even be *designated* except by their real names.

Just as the *Principles*, unlike the earlier work they draw on, no long pretend to pretend, so too the *Description* strips the fable from *L'Homme*:

> It is true that one may have difficulty in believing that the mere disposition of the organs should suffice to produce in us all the movements that are not determined by our thought; that is why I undertake here to prove it, and to explain the whole machine of our body so that we will have no more motive for thinking that it is our soul that excites in it those movements we do not experience as being conducted by our will, than to judge that in a clock there is a soul that makes it tell the hours. (*Descrip.* AT 11:246)

[10] See Ritchie 1964 in Moyal 1991, 4:289. Ritchie argues that even when the machine does have a mind, there is no seeing (292).

"The whole machine of *our* body": no need to pretend it isn't. The soul and its volitions will not be jeopardized. The real distinction is safely in hand. By 1648, moreover, it was no longer so radical as it might have seemed in 1633 to erase the vegetative soul. The long defense, with all its travails, of the tripartite soul was almost over. And finally Descartes has a explanation of generation that satisfies him. The man-machine comes, not from God, but from the seed, and that in turn from the parents—somehow.

In *L'Homme* the human remains prior to the machine, as that which God is to imitate in making the automatic "statues" of the new world. But in the *Description* the machine of our body imitates nothing; it is the original, not a copy. Even if the word 'machine' still reminds us of the artifact, the body is a machine only in the sense that mechanical causes—the "disposition of its organs"—suffice to explain all it does apart from the soul. The learned Anatomist disappears from the *Description*. In his place, an appeal to everyday knowledge: "there is no one who does not already have some acquaintance [*connaissance*] of the various parts of the human body, that is, who does not know that it is composed of a great number of bones" (AT 11: 226) and so forth; no one who has not seen animals slaughtered, and their interior parts laid open to view. "There is no need to have learned more Anatomy" than that to understand the *Description* (ibid.). One recognizes here the rhetoric of introductions. More than one thorny text has claimed to have no prerequisites. In fact Descartes promises to explain everything the reader needs to know as the occasion arises.

But perhaps there is more to the disappearance of the Anatomist than that. It is not some abstruse bit of knowledge that Descartes hopes to convey, but something basic, however radical. We tend to think that "because we have all experienced [*éprouvé*], since childhood, that some of [the body's] movements obey the will, which is one of the powers of the soul," the soul is the principle of all the movements of the body. Not knowing, furthermore, much of the body except its exterior, "we have not imagined that it has in it enough organs or springs to move itself in as many different ways as we see it move" (*Descrip.* AT 11:246). But of course it does, and because it does there is no need for a vegetative soul or vegetative powers.

To understand that is to refine the sense of everyday terms like 'heart' just as much as terms of art like *conarion*. The heart is *not* that part of the body whose specific nature is to have the purpose of forcing blood into the arteries. It does force blood into the arteries, but the addition of a purpose, or a power whose natural tendency is to do what the heart does is gratuitous once we understand not only how the blood circulates but also how the heart is formed. The *is like* of *L'Homme* can be dispensed with. From now on the object of Cartesian physiology *is* the body.

3.2 The Assumption of Normality

The second obstacle to stripping off the gloss of *usus* is that the fable of imitation hides an assumption of normality. The machine is supposed to imitate us as closely as possible in form and operation. Take the eye, for example. The machine's eye has a lens, a retina, vitreous and watery humors. Why? Because the eye does. But which eye? Some people's eyes have no lenses. Some retinas are detached. Some corneas are clouded by cataracts. Why not imitate them? The obvious answer is that the people God imitates are normal healthy people, not those deprived of vision, or people recently whacked on the head.

There is nothing wrong with Descartes' having supposed that the statues of *L'Homme* are statues of normal people. Physiology needs some sort of distinction between the normal and the pathological, even if it studies the pathological in order to understand the normal. But normality would reintroduce the finality that Descartes is doing his best to dismiss. It might be thought that Descartes could avoid the problem by limiting the scope of *L'Homme*. The statues imitate some but not all humans. But then the science of *L'Homme* would be the science, not of the species man, but of some individuals—which is to say, no science at all, if there is no principled basis for the selection. And that basis, it would seem, must be normality.

Certain passages in *L'Homme* that consider individual differences suggest another shift by which to elude the assumption of normality. In the brain, for example, the more feeble of the animal spirits that issue from the pineal gland tend to exit the chamber of the gland not by the pores that lead to nerves but through passages in the bottom of the chamber that lead to the nose and throat. If their way is blocked somewhat, they will eventually force their way out, and tickle the interior parts of the nose, "which causes *Sneezing*." If the passages are blocked even more, they will rise toward the top of the chamber, and cause fainting, vertigo, and "troubles of the imagination." (*L'Homme,* AT 11:172).

The alternative to imitating a normal person would be to show that the machine, with the structures one imputes to it, can simulate the whole range of phenomena that occur in humans, without distinction between the normal and the pathological. So Descartes explains, to take another example, that in a machine that imitates the body of a child, the matter of the body is "so soft, and its pores so easy to enlarge, that the parts of the blood that enter thus into the composition of the solid members [of the body], will typically be a little larger than the particles they replace," and so the soft machine will grow. But the matter of the body "will harden bit by bit" (by the operation of growth itself) and its pores will no longer admit particles larger than those already in them (*L'Homme,* AT 11:126). Implicit in the explanation of growth is a whole spectrum of machines, from the very soft to the very hard, whose behavior with respect to the accretion of matter to their parts is being explained.

To describe *all* the operations, normal or not, of the machine is a project worthy of the man who once thought he could deduce the positions of the fixed stars.[11] Simplify the project: take not a man-machine, but a piece of wood. In principle, by the laws of motion and using some hypothesis about its composition, one could explain every phenomenon it had a role in. But what sort of science would that be? The situation is worse with physiology. There one would have not one machine-type, but indefinitely many, to cover all the possible diseases, abnormalities, and accidents that befall us. You might say: yes, but some of them befall us only insofar as we are collections of bits of matter, and those events may be relegated to physics or mechanics or to the physical chemistry that Descartes projected in another letter of the same period (To Mersenne, 5 Apr 1632, AT 1:243). The difficulty is that there seems to be no principled way to make the required distinctions among the objects and events of one science and those of another, and thus no way to keep physiology from expanding until it becomes a science of everything.

A better answer is this. Physiology studies self-moving machines, *qua* self-moving. Its principal object is to study the means by which the machine moves itself, by which it replenishes the fuel that enables it to move itself, by which it finds the fuel, by which it maintains the mechanisms that enable it to find the fuel, and so forth. The language of means and ends here is only superficially teleological. What we are concerned with is efficient-causal relations. Physiology can also study the things that hinder self-motion and the rest. Again there is no supposition that self-motion is an end, but only efficient-causal reasoning.

Nevertheless, to choose self-motion as the ultimate effect of the machine, and to study all other phenomena involving the machine only insofar as they can be construed as helps or hindrances to it, remains arbitrary. The clock that runs and the clock that doesn't are on a par in the eyes of nature and nature's laws. Self-movers, in short, are not a natural kind in the Cartesian world. The choice to study them and those of their changes that pertain to them as self-

---

[11] To Mersenne, 10 May 1632, AT 1:250.

movers is not unlike the choice to study only those sounds that music has use for—sounds of definite pitch, for example, and among the infinitely many relations among them, those we call "consonances." It is not entirely implausible to think of physiology in that way, especially when if one treats it, as it was in fact sometimes treated in the Renaissance, as subordinate to the practical discipline of medicine.[12] We choose to study self-movers because *we* are self-movers, and to study them as physiology does is useful.

That position has the advantage of leaving untouched the basic apparatus of Cartesian natural philosophy. But it does so at the cost of transferring the science of life to the realm of the practical. That, it seems to me, goes against the spirit of *L'Homme,* and against Descartes' comments on his motives for introducing the fable of imitation. *L'Homme* is, textual lacunæ aside, continuous with *Le Monde,* part of the same project. When Descartes comments on it later, the motive he gives for taking the human machines to be already made by God in imitation of us is not that physiology is a practical science but that he had not collected enough observations and experiments on generation to explain how animals reproduce themselves. Nor did he have an explanation of their origins out of the initial state of the cosmos. The difference between physics and physiology is not as between a theoretical and a practical discipline, but as between a theoretical discipline near completion (so Descartes believed, at least) and one that had basic questions left to answer. When Descartes thought he had answers, he wrote the *Description*. Though practical aims do not go unmentioned in that work—the control of aging, for example—it is clear that it is intended to complete the unfinished business of the earlier work. The question of norms, therefore, remains open.

∞

Descartes had a very clear conception of what it would take to eliminate all appeals to powers, faculties, ends, and forms in physiology. The strategy is twofold. First, the prejudice by which his readers believe that all the actions of the body arise from the soul must be combatted. Second, once the force of that prejudice is abated, he must show that the resources of Cartesian physics are in principle sufficient to explain all those actions in which the will has no part. Since animals have no will, in them there is no residue left to psychology. For animals, physiology is all the science we need.

The second part of the strategy evokes from Descartes a whole battery of ostensibly mechanical constructions: the *feu sans lumière;* the self-regulating

---

[12] Siraisi 1990:79–80.

mechanism of antagonistic muscle action; the self-perpetuating cycles of the pulse; the replenishment of blood; hunger and satiety; the fluid mechanics of the spirits and their classification; the *Ur*-Animal; the blurring of the distinction between fluids and solids, and the leisurely flow of fibers; the prefiguring of function in the efficient causes of the circulatory system and the senses; the eversible organ of generation. If, as Deleuze and Guattari hold, philosophy is the proliferation of concepts, then Descartes was no less the philosopher in physiology than he was in metaphysics.

Too ingenious by half, said his successors, even as they adopted the fundamental basis of a mechanistic physiology. By the end of the seventeenth century, however, that basis itself began to tremble. *Life* began to seep back into physiology. The Aristotelian had had no difficulty distinguishing living from nonliving, even if the definition of life itself proved elusive. For Aristotelians it is a basic truth about the nonhuman world that some of the things in it have the power to grow, to nourish themselves, to reproduce, to sense, and so on, a truth not reducible to truths about the properties of matter considered apart from life. No temperament will supplant even the vegetative soul, and *a fortiori* none will suffice to explain the actions of higher animals.

Descartes, on the other hand, made the proof of concept, as I have called it, his main weapon against the vegetative and sensitive souls, partly in the interest of liberating the human soul from any suspect relation to matter, partly in the interest of erasing from natural philosophy the last and strongest bastion of forms and powers. He thereby—knowingly—eliminated life, not only from the world of *res extensæ,* but from the world of *res cogitantes* as well. Although even angels and God—immaterial entities like the soul—had long been said to live, there was, even among certain Aristotelians, an increasing reluctance to ascribe life unequivocally to spiritual and material substances.[13] It is as if the tension within the tripartite human soul, in which the noblest and humblest powers of living creatures were bound together, were at last proving too great. Descartes finally cut the strings. They were joined again, in the panpsychisms of Leibniz and Spinoza, in the panvitalism of La Mettrie and Diderot. But by then the Aristotelian soul was dead.

---

[13] Cf. Arriaga's distinction between 'physical' and 'intentional' life (Des Chene 2000, §2.1).

PART TWO

MACHINES, MECHANISMS,
BODIES, ORGANS

∞

[. . .] props, columns, beams, bastions, teguments, wedges, levers, fulcrums, pulleys, lines, vises, bellows, sieves, filters, channels, buckets, reservoirs [. . .]

—Boerhaave, *De usu*

Statues that walk, artificial Flies that buzz and take wing; Spiders of the same manufacture that spin their webs; Birds that sing, a golden Head that speaks, a Pan that plays the flute.

—Bouillier, *L'Âme des bêtes*

Perhaps he made one, perhaps not: "A statue, with some iron in its head and feet." Other bits of iron, imbued with *vis magnetica,* above and below. In its hands "a staff held in the manner of a tightrope-walker," with a string stretched within. When it is touched this automaton will, under the force of the magnets, move its feet, *spontè:* to the beat of instruments, for example.[1] So writes Descartes in an early fragment. His commentator Nicolas Poisson, in some need of establishing Descartes' credentials as an experimental philosopher, thought that Descartes actually made the statue, "to verify by experience what he thought about the souls of animals." But Descartes writes *ponatur,* 'let there be made,' not 'I have made.' It is the 'let' of 'let a circle be

---

[1] "Cogitationes privatæ," AT 10:231. The "Cogitationes" (whose title was probably added by Foucher de Careil) are dated 1619–1620, and exist only in fragments copied for Leibniz in 1675–1676.

given,' the 'let' with which an object is ushered onto the stage of imagination. Poisson may well have been inventing a little himself.[2]

It hardly matters. In the science of life the machine is a conceit, a metaphysical notion.[3] It was commonplace when invoking the machine to enlarge upon the inability of human industry ever to emulate the minute, myriad springs and levers in God's devices: the point was not to build, but to understand. Of the automata named time and again in lists like Boullier's, few ever existed. But the uses to which they are put do not require that Albert the Great should actually have made a talking head, or that Archytas' dove could fly. Only their *ideas* need be admitted, only their existence in the shop-windows of the possible.[4]

But what were those ideas? What is the "machine" so frequently invoked in the science of life? And to what purposes is it put? It would be naïve to expect those questions each to have but a single answer. 'Machine,' like other concepts in its neighborhood—nature and art, the organic and the inert, animate and inanimate—is better thought of as a variable, or a zone of variation impinged on by others, than as an essence. The historian can arrest that variation at a moment, in some chosen corpus of texts, or follow it through periods when it seems visibly to have shifted its center in conjunction with, or in opposition to, its neighbors, and describe the regions it traverses in their company. But it is unlikely that those regions will have sharp boundaries, or that the concept will admit a precise, sempiternal definition specific enough to be of use in understanding particular events.

At least one shift has occurred since Descartes, and should put us on our guard. G. E. Stahl, writing on the "differences between mechanism and organism," describes a river, its course determined by the "accidents of terrain"; around it there are forests abundant in game, and plains "naturally adorned with agreeable sites." "No one," Stahl concludes, "will deny that there we

---

[2] See Rodis-Lewis 1971:89, 472n71, and Poisson *Commentaire* pt5, Observation 3, p156.

[3] "L'horloge qui sert de modèle logique aux conceptions mécanicistes de l'univers et de la société est, bien entendu, un concept purement métaphysique" (Schlanger 1971:52; she credits Deutsch 1951:234). I have learned much about the topics of this chapter from Schlanger's neglected work.

[4] One point on which Salomon Caus promotes his *Raisons des forces mouvantes* is that "although it is true that Jacob Besson, Augustin Ramelly, and certain others have exposed certain Machines invented by them on paper, few of them could have any effect, and they have believed that by a multiplication of gears, their machines would have an effect" (Caus *Forces mouvantes*, "Epistre au Lecteur"). Michael Segre, agreeing with Caus, argues that Ramelli's machines belong to a "recreational mathematics" analogous to the "recreational mathematics" found in authors like Leurechon (*Récréation*, 1624; see Segre 1994) and Mersenne. Caus himself offers an example of the genre—a system of six gears with a combined ratio of $12^6 = 2,985,984$ (Caus *Forces mouvantes*, Thm. 16; see Fig. 6).

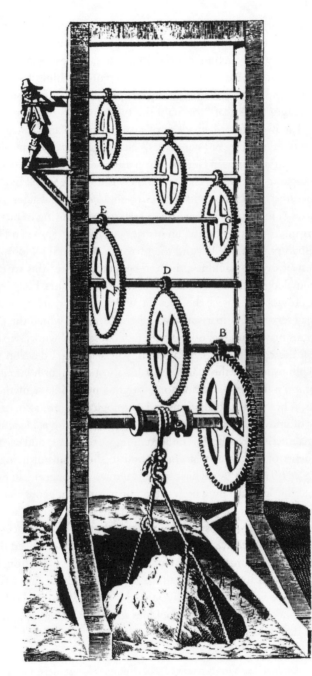

Figure 6: Recreational mechanics (Caus *Forces mouvantes*, Thm. 16).

find a genuine, real mechanism." But when "human industry, with intentions and for purposes that are social and arbitrary, masters to its liking this docile element, [. . .] by conducting it though channels, enclosing it in reservoirs and ponds, to which it has been brought by certain machines [. . .]," then the river and its environs "take on an organic character." Schlanger, summarizing the passage, notes that "for Stahl the mechanical is the fortuitous, the organic is what is organized in view of an end; organ and instrument, in agreement with etymology, are synonyms" (Schlanger 1971 : 50, quoting Stahl *Œuvres* 2 : 292–295).[5] It was in German Romanticism, she argues, that the contrast of the organic and the mechanical took on the value it still has, and the *mechanical* became coincident with the *inert,* the *lifeless* (Schlanger 1971:51).

In Descartes, a machine is indeed not a fortuitous gathering of bits of matter. It is "organic," in Stahl's sense, an instrument made to some end by us or God. But God's purposes lie beyond the province of natural philosophy. The purposiveness of divine machines, the animals and plants of this world, is no sooner acknowledged than set aside. Instead, the machines built by us are thought to provide an object the idea of which is transparent, and which therefore may serve in comparisons as a schema through which the causes of vital operations may be understood. There is, in Descartes' works, virtually no reflection on the concept of machine itself. In Chapter 4 I develop some of the absent "philosophy of the machine." One must distinguish *mechanism,* or the "mechanical philosophy," a more or less tight-knit bundle of methodological precepts, schemes of explanation, and ontological claims, associated with an ideology of knowledge and control through making; the *machine* as a concept and imaginative scheme for understanding organisms; and *mechanisms,* the components of machines that correspond to the organs of living things.

The *Traité de L'Homme,* as we have seen, is framed as a conceit, an extended comparison. The standard by which the comparison is judged is that of successful *simulation.* God is supposed to have made statues as much like us in the disposition of their parts as possible, statues that, but for the absence in them of reason, might well deceive us as the paintings of Apelles deceived their onlookers. Machines of human manufacture can be vehicles of illusion, making signs in the sky, for example, to mislead the credulous. Yet the simulation proposed in *L'Homme* is a means of conveying the truth about the body. There is, in the end, no illusion. Divine machines imitate nothing: they *are* us, in every action that does not require thought.

---

[5] Leibniz, by contrast, treats the two terms as coextensive: "an organism is formally nothing other than a mechanism, even if it is more exquisite and divine" (Giglioni 1995:258).

Still Descartes cannot, even when the scaffolding of fable is dropped in later works, entirely avoid the appeal to simulation. In Chapter 5 I examine the to-and-fro among simulation, artifice, and illusion. Their primary role in the eventual science of life is indirect. Invoking them serves primarily to assist us in letting go the prejudices which everyday life has inculcated in us. But the necessity for Descartes to combat prejudices and at the same time develop the philosophy that is supposed to replace them introduces a systematic ambiguity into his exposition, an ambiguity affecting the claim, for example, that animals live, or that they have souls. When confronting his opponents Descartes will insist that animals do *not* have souls, if by 'soul' is meant what the Aristotelian means. On the other hand, he will say: you may, if you wish, continue using the word 'soul' of something in animals—but then you must consider that the soul is only the blood, or the heat of the heart, and nothing more. Though philosophers occasionally misunderstand the tactic, and take Descartes to have contradicted himself, in context there is no confusion.

We have seen in Chapter 1 that the animal-machine enjoys a kind of autonomy. Indeed, were it not for the need to replenish the blood, the machine, for all one can tell, might operate on its own indefinitely, or until the petrifaction of age brought it to a standstill. Considered in abstraction from its maker, the animal-machine and its mechanisms have, it would seem, only a specious unity, either that which is projected onto it by us when we treat them as if they were human artifacts, or the unity conferred on them if we mistakenly attribute to them souls resembling our own. Only in relation to divine ends is the animal-machine *one* thing, and therefore, to borrow Leibniz's turn of phrase, one *thing*. Yet Descartes insists that divine ends, except those that God has chosen to reveal to us, are incomprehensible. The organism of common sense disappears, replaced by an instrument whose parts are united only by something extrinsic to them, like a crowd momentarily transfixed by some spectacular event.

The *Passions* offer, not without hedging, a notion of unity that would recuperate the organism, the mutual dependence of its parts: "when one part is taken away, the whole body is made defective." The body is "in a certain manner indivisible," and thus the soul cannot be joined to any one of its parts to the exclusion of the rest (*PA* 1§30, AT 11:351). Rather than settle the question of unity, however, this brief passage raises further questions: how it is to be fitted with Descartes' properly physiological works, how the condition of one part of the body might depend on another, and whether the notion of defect would make sense in any animal but the human. Those questions will occupy Chapter 6.

[4]

Tools of Knowledge

The machine is conceptually complex. Ontologically, it is a thing whose properties are just those admitted into the "mechanical philosophy" of nature: for Descartes, these are the modes of extension. Cognitively, it is a schema, a means of imagining the operations of things. Aesthetically, its more intricate forms provide occasion for admiration, for wonder, whether at the handiwork of the skilled craftsman and designer or at the exceeding wisdom of God— but also the means by which humans, if not God, can make something appear to be what it is not, or nothing appear to be something. In this chapter I consider the first two aspects of the machine, and in Chapter 5 the third.

In what follows I make a threefold distinction between mechanism (or the "mechanical philosophy"), machines, and mechanisms. About the mechanical philosophy not much needs to be said here. In application to living things it amounts to two presuppositions.

(i) The properties of matter are all "mechanical," whether by 'mechanical' one means with Descartes only the modes of extension, or with other philosophers those modes and a small list of further properties not reducible to them, like impenetrability. The key point is that in the present context none of the properties or entities attributed to animals by Aristotelians or Galenists, which like vital heat or the vegetative soul were thought to belong *only* to living things, are among the mechanical properties of matter.

(ii) The operations of things composed of mechanical matter are to be explained in principle according to laws formulated in mechanical terms. In practice, that often amounts to comparing things with machines whose operations seem obviously to be explicable by such laws. The point is not that one hopes to write down a grotesque set of equations describing the motion of a dog's tail as a function of the irradiation of its retina. It is that, on the supposition that a dog is a machine, there *must be* an explanation which adverts only to mechanically described parts and causal relations—in Descartes' physics, these are bits of extended matter and collisions of particles. The task of persuasion lies not in bringing the laws of nature to bear directly on living things, but in making good the supposition that living things are machines.

'Mechanism' thus signifies a cluster of claims about the nature of corporeal substance and the character of natural change; and also certain doctrines about explanation, together with the practices of explanation that those doctrines were meant to codify and promote.

As for the machine and its mechanisms, the machine is a whole of which mechanisms are parts. The machine has ends and capacities as a whole, its mechanisms have ends and capacities as parts. Accompanying that distinction is a method—the analysis of capacities, in which the division of the body into systems and organs is explicated by assigning to each a specific end or function, and demonstrating from its structure or "disposition" that the system or organ can accomplish its end or fulfill its function. In physiology, machines correspond to organisms, and mechanisms to organs.[1]

4.1 Exhibits

Consider first the machine as schema. What could be easier to imagine than this? "While this gland [the pineal gland] is kept leaning to one side or another, its leaning prevents it from being able to receive as easily the ideas of objects that act on the organs of the other senses. Here, for example, while almost all the spirits produced by gland H issue from the points *a, b, c,* not enough of them issue from the point *d* to form the idea of the object D, whose action I am supposing to be neither as lively nor as strong as that of ABC. From which you see how ideas hinder one another, and how it happens that one cannot be very attentive to several things at once" (*L'Homme* AT 11:185; see Figure 7).[2] And indeed we can, for there on the page is a man-

---

[1] On the analysis of capacities in cognitive psychology and biology, see Cummins 1983.

[2] In *L'Homme*, the term 'idea' refers to the impressions made on the pineal gland or the interior of the brain through the action of the nerves or the gland itself (AT 11:176–177). On this use of 'idea,' see Michael & Michael 1989; Baker & Morris 1996:79n50.

Figure 7: How one object distracts us from another (*L'Homme*, AT 11, Fig. 35).

machine, whose gaze, fixed on an arrow, prevents it from smelling the flower beneath its nose. On its gland H we see the point *d* and the animal spirits issuing from it, overpowered by the animal spirits issuing from *a, b,* and *c.*

The *Traité de l'Homme* continually solicits the reader's attention and assent: "It will be easy for you to believe . . . ," "Consider that . . . Consider also . . . After that, consider . . . Moreover, consider . . . ," "I want to make you conceive . . . ," "But I have not yet shown you how . . . ," "I desire again that you should reflect a little on what I have just said about this machine" (183, 186–187, 189, 194, 200). The density of such phrases is remarkable. It is the voice of the purveyor of marvels, of what the eighteenth century would call a "projector" exhibiting his latest invention.[3]

The machine is visible, transparently articulated in a way that forms and faculties are not. It can be *shown.* In scores of books that begin in print with

[3] The reader is implied at least once to be viewing the machine not just from outside, like the implied spectator of Caus's plates, but from *within.* The threads that run through the nerves, Descartes says, "communicate easily from one end to the other, without being hindered by the twists and turns of the paths through which they pass." And he continues: "But in order that these twists and turns should not hinder you from seeing clearly [. . .] observe in the figure here the little threads [. . .]" (*L'Homme*, AT 11:174). This after the reader has been invited to contemplate the actions of the spirits issuing from the pineal gland resting in its chamber in the brain. It is not hard to believe that one is supposed to be inside that chamber, looking, or trying to look, out the pores that line its walls. It is as if the reader were momentarily asked to occupy the place of the gland itself, to take the pilot's seat.

Roberto Valturio's *De re militari* (1472) and continue through the *Encyclopédie* and its famous *Recueils de planches* (1762–1777), the machine *was* shown, its occluding surfaces removed to show its insides, its key components enlarged, or disassembled in an "exploded view," already found in Leonardo. The *theatrum* or theatre, whose etymology links it with beholding, wondering, and contemplation, is, as many titles of the period indicate, the natural place of the machine.[4]

It is, I should add, a theater of the mind—aided by the hand of the artist. Descartes advised Mersenne to interpret his badly-drawn figures according to the text, since otherwise the engraver would not make sense of them. This Mersenne did well enough that Huygens, evidently a connoisseur, expressed his pleasure in them.[5] Florent Schuyl, producing figures for the 1662 Latin translation of *L'Homme*, "did not merely content himself with accommodating the text as well as it could be; he wanted [the figures] to be as beautiful as possible, and one feels in them the hand of a true artist." When Clerselier published the French version, he found Schuyl's figures to be better than the ones he had used, "considering only the engraving and the printing," but less suited to the understanding of the text. He used instead figures provided to him independently by Louis de la Forge and Gérard van Gutschoven, sometimes printing both (AT 11:vi–vii, xi–xiii). As we have seen, the figures differ significantly in detail. But what matters is that the operations and structures of the body be not merely described but depicted—well, if possible, but somehow if not.

Description and depiction were not independent. The passage on the pineal gland, and passages like it in Descartes' natural philosophical texts (for example, the description of the action of antagonistic muscles: see §1.1, Fig. 2 above) are almost unintelligible except by consulting the figure, with its numbered and lettered elements, as one reads. We are so used to back-and-forth reading of this sort that we hardly notice it. Yet the *De Anima* commentaries of Toletus, Suárez, and the rest are almost bereft of figures. Substantial forms

---

[4] See, for example, *Cult. delle Macchine* 1989. On theatres (dramatic and anatomical) see Cavaillé 1991, c. 1, and Sawday 1995, c.4.

[5] Descartes to Mersenne, March 1636; AT 1:339; Huygens to Descartes 5 Jan 1637; AT 1:346. Huygens adds that he would have wanted a paper of higher luster, and wider margins. Earlier, proposing to publish the *Dioptrique*, he had suggested that it be illustrated with woodcuts, not copperplate engravings, and with the illustrations spread throughout, and not gathered at the end, since that arrangement would impose on the reader "la peine de l'oiseau, qu'on dit travailler a percer les arbres, & en faire tant de fois le tour, pour veoir s'il a passé" (Huygens to Descartes 28 Oct 1635; AT 1:325–326). Though one might suppose that even if they had not been printed with such care, Descartes' works would have been as influential, still the manner of presentation, including the quality of the materials, was not (nor is it now) irrelevant to readers' response.

and active powers, of course, do not beckon the engraver. But even when it would be appropriate, as in anatomical descriptions, the texts rarely include figures, nor are figures needed to understand them.[6]

The theatre of the machine is a theatre with *captions*. The caption supplies two things that a picture can at best put us in mind of: purpose and motion. In Salomon de Caus's *Raisons des forces mouvantes,* the titles of the "problems" of the First Book, with few exceptions, state an end: "to counterfeit the voice of small birds by means of water and air," "to make a machine that will move by itself," "to make an admirable machine, which being set at the foot of a figure, will emit a sound at sunrise, or when the Sun shines from above, in such a way that it will seem that the figure makes the sound."[7] The 'machine' and the 'for what' of finality are inseparable in works like that of Caus. Even if the purpose of a machine is frivolous, even if it was never intended that the machine be built, still it must have one. I will return to the relation of machines to ends in §6.1. Here it suffices to note that the 'for what' is not in the picture. It is invisible, not because we lack eyes to see it, but because it exists only in the caption. We must be told; we cannot see.

Nor can we see the motions by which the machine fulfills its purpose, or the forces that cause those motions. Problem 9 in the First Book of Caus's work describes a "vessel, which if water is forced into it, [the water] will then issue forth with great violence" (see Fig. 8):

Let there be a vessel of copper, quite round and having sufficient force to sustain the effort of the air, and let it be well sealed and soldered on all sides; after which two pipes, namely AB and CD, must be soldered to it, in such a way that each of the ends [of the pipes] inside the vessel approaches the bottom as needed in order to allow water to pass into it, and for each of the pipes there will be a key or faucet to restrain the water when it is inside, being put inside with a Syringe through the pipe CD. The end of the Syringe must well be adjusted to the end C of the pipe, in order that water forced into the pipe should not come out again at the joint, and at the instant that the water is forced inside one must open the key G and then close it as soon as there is more water in the Syringe, and thus when one wishes to make the water issue forth, one turns the key or faucet F and then it will issue forth through the pipe A (whose mouth is no bigger than

---

[6] Compare, for example, Toletus's description of the eye with Descartes' (Toletus *In de An.* 2c6q16, *Opera* 3:83v–85r, and Descartes *L'Homme*, AT 11:152–155; *Diop.* 5, AT 6:115–123). Of course the use of figures in optical and mechanical treatises was not new. See, for example, the plates in Lindberg 1970.

[7] These are the titles of Problemes 10, 12, and 35.

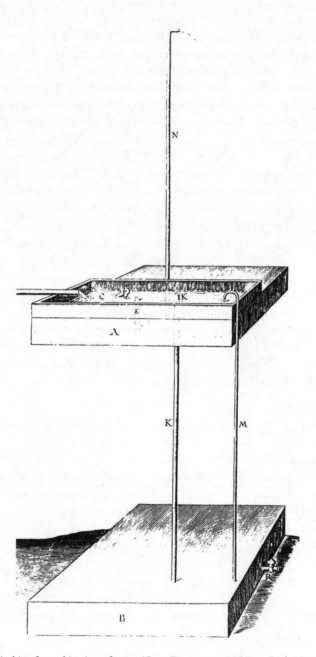

Figure 8: Machine for making jets of water (Caus *Forces mouvantes* Liv. 1, Prob. 9).

the size of a needle) to a height of twelve or fifteen feet, which will be a pleasant sight. (Caus *Forces mouvantes*, Liv 1, Prob 9)

We see the syringe with its plunger, the pipes and their faucets, the copper vessel. But we do not see the water or the jet issuing from A. In other figures, we do see water, and in a few, ripples or waves suggest motion, though typically not the motion that does the work. Nevertheless the course of events remains, as it does in the figures of *L'Homme*, for the caption to describe.

The exhibit of a machine in Caus's work has three components: a title, stating its purpose; a picture, exhibiting its parts and their spatial relations, or what in the period was called its disposition; and finally a narrative, often having the form of a procedure by which the machine can be made to fulfill its purpose.[8] The procedure describes a linked sequence of events (branching narratives are rare, as are confluent narratives in which two sequences merge into one). The picture and the objects referred to in the narrative are made to correspond by letters and numbers, whose order quite often reflects the order in which the objects are set in motion.

The book too is a kind of machine, whose parts "move" when it is read.[9] Nor does the construction of exhibits end with the printed page. Mersenne's copy of his *Harmonie universelle* is peppered with revised figures, annotations, new sketches (Mersenne *Harm. univ.* 1: 96, 157, 166, 189, etc.). Material crafts of engraving and printing, imaginative skills of drawing and seeing, and the intellectual grasp of final and efficient causal relations join forces in the act of

---

[8] *Dispositio* is defined in Aristotle's *Categories* as a transient *habitus* or state (1c8, 8b25); in the *Metaphysics* it is defined as "the order of that which has parts, according to place, *potentia*, or form" (5c19, 1022b1). 'Order' typically means 'order with respect to a scale,' quite often a scale of perfection or power (Des Chene 1996:151–152). There is a discussion of the term in Carter 1983:99–101; but I should note that Descartes' use was not at all "strained" in the period, and that I am not sure that a clear-cut use of the term in the sense of "stance or mood" is to be found in Descartes (Carter cites no texts). The "primary meanings" of these words, as Carter calls them, are less relevant to understanding the texts of the period than their *defined* meanings (and the usage those definitions are intended to reflect).

[9] The comparison is not far-fetched. Charles-Alphonse Du Fresnoy, in his *De arte graphica*, writes that after choosing a subject, "I obtain first of all a bare canvas on which one must arrange the whole machine (so to speak) of the painting." His translator Roger de Piles comments: "It is not without reason that our author uses the word *machine*. A machine is an exact assemblage of parts of which one must foresee the accord and appropriateness in producing a beautiful effect" (Du Fresnoy *De arte*, quoted in Becq 1982:274). In his own *Conversations sur la peinture*, Roger de Piles writes, "The painting may be regarded as a machine whose pieces must exist for one another and produce together but one effect" (Piles *Conversations* 297, quoted in Becq 1982:275). The painting is a machine for pleasure, the pleasure we take in harmonious arrangement suited to its subject.

reading the exhibits of Caus and Descartes. The purpose is to make the reader see, imagine, conceive—*faire voir, faire imaginer, faire penser.*

The historian too is moved to imagine and conceive. But unlike seventeenth-century readers, the historian does not read Descartes and Mersenne in order to learn the latest state of scientific thought. That part of their intention has lapsed. It is not surprising that the rhetorical aspects of the text—its "literary technologies"—should then become more prominent. They were, of course, never absent. Polished prose or copiousness of learning are certainly not incidental to the effects of a text—including that of conveying knowledge. But philosophical works were supposed primarily to act on the intellect. Their action on the passions was, in principle, subordinated to that end. To dwell on rhetoric alone is to lose what for Descartes' early readers was foremost in their expectations and their experience of the text: comprehension of the experiments, hypotheses, and so forth that its words and images were meant to convey. In books on machines, the aim was to convince the reader that the device depicted and described could—if only in the reader's imagination—carry out its assigned purpose.

4.2 The Analysis of Capacities

The structure of the three-part exhibit is reflected in the definition of 'machine' taken by Caus from the tenth book of Vitruvius' *De Architectura,* one of the *loci classici* of machine theory.[10]

> The word machine [. . .] signifies an assemblage and firm conjunction of timber-work or some other material, having force and movement either from itself or by some means or other, and there are three kinds: the first kind the Greeks call Acrobactic, and is that which serves to raise all sorts of loads [. . .]; the second kind is called Pneumatic, and acquires movement by water and air, and in which there are various machines, serving to decorate grottos and fountains; the third the Greeks call Banauson, and [it] serves to raise, pull, and carry

[10] Although Vitruvius' definition was often adapted by early modern authors, the difficulty of defining 'machine' is a recurrent topos in the theory of machines. In his *Theoretische Kinematik,* one of the most successful nineteenth century works on the theory of machines, Reuleaux, giving seventeen definitions, says that they "show how uncertain, and often how altogether indefinite, have been the attempts made to define the machine even by those who must have known the thing itself." He of course adds one of his own. His translator Kennedy lists seven more (Reuleaux 1874–5/1876:585–586).

from one place to another all sorts of loads, and likewise as a force to do a number of things difficult for us without this help.[11]

The assemblage is what can be pictured; "force and movement" are the object of the narrative; and the purpose is to move loads. Even pneumatic machines move water, though for pleasure, not use.

It is not difficult to see here the four Aristotelian causes, and thus simply an application of the usual scheme for explaining natural change. In fact nothing distinguishes the machine, so defined, from living things or their organs. Suárez writes, for example, that the first cause of the voice of animals is the soul, in whose service there are five "instruments": the lung, the trachea, the larynx, the glottis, and the palate. From these five parts is "composed [*aggregatur*] the entire instrument for the formation of the voice, which is similar to a pipe [*fistula*, 'pipe' or 'panpipe'] [. . . .] For the trachea [*oblonga arteria*] is like a pipe, the lung corresponds to a mouth putting air into it, the glottis of the larynx corresponds to an opening at the top part of the pipe, to which the hyloid bone [*os canens*] is applied, and as the air enters it, and the holes are closed or opened, the sound varies [. . . .] By the instrument described, the voice is formed."[12]

What then distinguishes the machine? Three candidates suggest themselves: (i) that the machine is an "assemblage" or "aggregate" of distinct functional parts, linked together by their mutual action; (ii) that it is an instrument; (iii) that it is artificial, or at least that the paradigmatic machine is an artifact. We needn't choose among them. All three contribute to the concept, and to its usefulness as a schema for understanding animals. The first opens the way to an *analysis* of capacities; the second to placing the machine in a causal *order;* the third to determining its *nature,* and in Descartes to the elimination of Aristotelian form from the nonhuman animate world.

---

[11] "Ce mot de machine, comme dit Vitruue signifie [*postil*: Vitruue livre X. Chap. I.] un' assemblage & ferme conionction de charpenterie, ou autre materiel, ayant force & mouuement, soit de soymesme, où par quelque moyen que ce soit, & y en a trois genres: l'une appellee des Grecs Acrobactique, & est celle qui sert à monter toutes sortes de fardeaux en haut, dont se seruent les Charpentiers & Massons, & mesmement les Marchands, à tirer toues sortes de machandises hors des Nauires, le second genre est dit Pneumatique, lequel acquiert mouuement par leau & l'air, dont il y a diuerses machines, seruantes à la decoration de grotes & fontaines, le troisiesme est dit des Grecs banauson qui sert a esleuer tirer & porter de lieu à autre toutes sortes de fardeaux, & mesmement à seruir de force à faire plusieurs choses à nous dificilles sans cest aide" (Caus *Forces mouvantes*, "Epistre au Lecteur").

[12] Suárez *De An.* 3c21§1, *Opera* 3:679–680; Suárez's references are Galen *De usu part.* 7c4–5, *Opera* 3:521–528, and Vesalius *De fabrica* 2c21.

1. Details and Components.

Pictures of machines sometimes contain details of key parts. A figure in Vittorio Zonca's *Novo teatro,* for example, shows a winch whose most noteworthy feature is an endless screw [see Fig. 9, (i)]. We see the whole machine, and above it, at a different angle and on a slightly larger scale, the screw. In front of and behind the winch other parts are drawn, illustrating the joints that in the assembled machine are hidden.[13] In Caus's work a machine "for raising water from a spring or river by the force of horses" is first pictured in its entirety, horses and all [see Fig. 9, (ii)]. The next plate [see Fig. 9, (iii)], a "plan d'ortographie," shows the heart of the machine: two pistons that go up and down alternately. In that plate an inset shows the crucial, and most ingenious, parts of all, the two *seaux* or buckets that lift the water out of the pistons.[14]

The detail magnifies, making the construction easier to see. More importantly, it isolates. It shows a genuine *component,* as opposed to an arbitrary *part,* of the machine, typically a working part that figures in the narrative description of its operation. It is a component because it has a delineable function. The screw multiplies the force applied to the crank; incidentally changing its direction.[15] The bucket, evidently, scoops up water, but it must also be capable of rising up and down in the cylinder: hence the valve, labelled C. In Caus's *plan* (Fig. 9, (iii)), the gears, pistons, cylinders, and buckets are isolated; they form a *system* or *mechanism,* a machine within the machine. We are to understand that the force produced by the horses has been transmitted somehow to the axle marked *d.* Caus's *plan* shows what happens next.

The detail abstracts to some degree, as engineering drawings do, from concrete realizations of the machine. Although Caus specifies that the gears have nine teeth, and that the *seaux* travel three feet (98cm or almost 39in), it is the geometry of the linkages that matters, not the absolute sizes of the parts or even some of their relative dimensions. The detail abstracts also from the causal chain in which it is embedded. We don't see the hand that turns the crank, or the horses that supply power to the pump; nor do we see the gear

---

[13] Zonca *Novo Teatro.*

[14] Caus *Forces mouvantes* Liv 1, Prob 3 & 4.

[15] The problem of determining the ratio or *raison* of the force of the worm gear (*pignon à vis*) is reduced by Caus to that of the ratio of the force of an ordinary gear, and that in turn to a basic theorem about the balance beam: the movement of each weight is proportional to its distance from the center of gravity. For gears, the theorem yields the conclusion that the ratio of forces is proportional to the number of times the smaller gear turns for each turn of the larger (and thus to the ratio in the numbers of teeth). In worm gears, one turn of the crank advances the wheel one tooth, and so the ratio is simply the number of teeth on the wheel. (Caus *Forces mouvantes,* Theoresmes 11, 12, 15, 17; cf. the "Explication des engins" in Descartes to Huygens, 5 Oct 1637; AT 1:441–442).

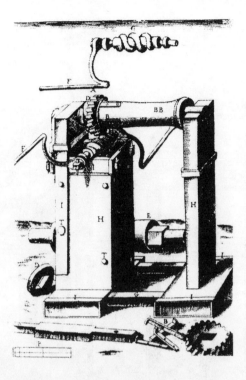

Figure 9: (i) Winch with endless screw (Zonca).

that meshes with the screw, or the trough into which the water flows. The component is lifted out of that particular sequence of causes and effects, and can be inserted into others, with different antecedents and successors.

We see the analysis in practice in the machines themselves, in the repeated use, for instance, of linkages that convert rotations into linear or reciprocating motions. We see it in theory in Caus's preliminary theorems, where a construction credited to Hero of Alexandria is explained, and then applied in various problems about the lifting of water.[16] Hero's device transforms the force of gravity into pressure, which forces water upward out of the sealed container A (see Fig. 10).[17]

---

[16] Caus's immediate source for the device is Cardano *De subt.* 1, *Opera* 3:365.

[17] Water poured into the open basin *C* tends to move through the pipe *CD* into the lower closed reservoir *B*. That compresses the air above the water in *B*, and the pressure is conveyed to the upper closed reservoir *A* by the pipe *EF*. The pressure of the air in *A* forces water to exit through the pipe *HL*. Eventually *A* will need to be refilled using the faucet *I*, and *B* will need to be emptied using the faucet *B*. Caus offers a calculation of the height to which water will rise when forced out of *A*.

Figure 9: (ii) Two-cylinder pump (Caus).

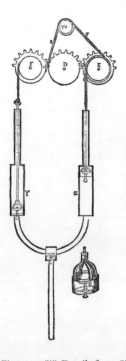

Figure 9: (iii) Details from (ii).

One disadvantage of Hero's device—that water must be added periodically to the open container *GC* on top of the sealed container *A*—remains. That deficiency is made good in the next problem, which combines the device of Hero with a mechanism for opening and closing two faucets automatically; part of that mechanism is a bucket that empties itself periodically, a device reused in several other problems.

2. Simple Machines

Caus's work and others like it reflect what presumably began as a kind of implicit craft knowledge, a lexicon of reusable mechanisms. Some of that knowledge had already been codified by the Greeks into the doctrine of simple machines, in which the forces of six devices—the pulley, inclined plane, wedge, wheel, screw, and lever—are reduced to a single proposition concerning the balance beam.

Those devices "can be applied," Descartes writes in his "Explication des engins," "in an infinity of diverse ways" (Descartes to Huygens, 5 Oct 1637; AT 1:435–436, 447). But the engineers' lexicon contained many more devices, whose actions, though sometimes reducible to those of simple machines, were nevertheless worth isolating and reusing in their own right: the two-cylinder pump, the self-emptying bucket.

Commentators on the "Explication" have dwelt on its quantitative aspects, especially the so-called "principle of virtual work" of which the proposition about the balance beam is a special case. But the mechanical analysis of capacities is quite independent of any quantitative assessment of forces and motions, quite independent, even, of the notion of law. Descartes, for all that he appealed to machines in his physiology, never gave a quantitative description of anything more complex than a worm gear.[18] But he understood the

---

[18] The *Dioptrique* describes a machine for grinding hyperbolic lenses based on the proposition that the hyperbola is a conic section (*Diop.* 10, AT 6:216–224). But the analysis is directed at showing how a particular trajectory—that of the grinding element—is to be determined by the constrained motion of the element. It has nothing to do with force or work.

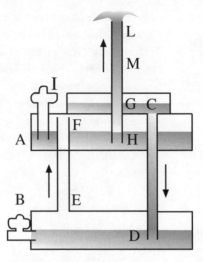

Figure 10: Device for lifting water (adapted from Caus *Forces mouvantes*, Thm. 6).

principles of mechanical analysis, and especially those of "automatic" machines, like Caus's adaptation of Hero's device (*Forces mouvantes* Liv. 1, Prob 6). We have already seen those principles at work. The heart is isolated from the rest of the body, and the action of the blood within it studied quite apart from the source and destination of the blood that flows through it. So too a single pair of antagonistic muscles is extracted from the instance at hand (the muscles of the eye) and subjected to analysis in terms of bladders and valves. Sometimes the extraction is literal, as in Descartes' experiments with the eel's heart. But the crucial step is one of conception—the screening off of antecedent causes and subsequent effects.

The parallel with the illustrations of *L'Homme* is most striking in the series representing the brain and pineal gland.[19] We first see the brain in a figure depicting the upper and lower extremities of the basic circulatory system [Fig. 11 (i)]. We next see two figures of the whole brain, from above and from the side, and then the bottom part, [Figs. 11, (ii); (iii)] with the pineal gland, the nerves, and the pores at the end of the nerves. Finally we see the gland isolated, as if sitting on a perforated rug [Fig. 11, (iv)]. If the object near *C* in the first figure is the gland (and if Adam-Tannery has reproduced the figures to scale), then the gland in the last figure is magnified by ten or twelve times. The succession moves from the whole circulatory system, to the brain, then to the subsystem Descartes uses to explain global states like sleeping and

---

[19] On illustrations in Descartes' works, see Baigrie 1996a.

waking, and the process of sensation, and finally to the central mechanism—the gland
floating above the threads that anchor it to the
brain, and directing animal spirits toward the
pores that surround it.

The cumulative effect is powerful even
when one knows that the anatomy is, in
Alquié's words, *un peu fantaisiste,* and that the
pineal gland has no direct role at all in cognition. Powerful in just the way that a *roman
philosophique* ought to be: we have an explanation, complete to the smallest detail, of all
the operations once attributed to the sensitive soul. The illustrations, moreover, have
drawn us into the brain itself, close enough
to see to gland leaning this way and that. Figures containing details from larger structures
were a staple of anatomy books. In his *Theatrum Anatomicum* Caspar Bauhin includes a
drawing of the brain stem in which the gland
can be seen sitting just behind the so-called
"testicles" of the brain, not at all prominent

Figure 11: (i) Circulatory system.

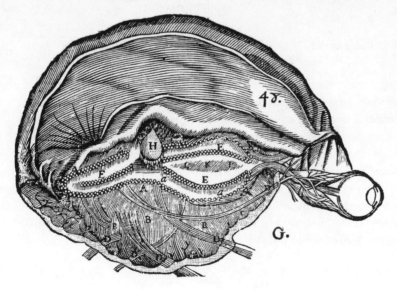

Figure 11: (ii) The brain (side view).

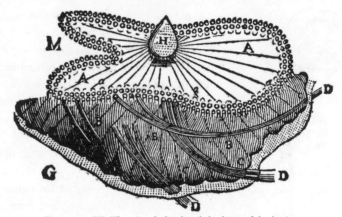

Figure 11: (iii) The pineal gland and the base of the brain.

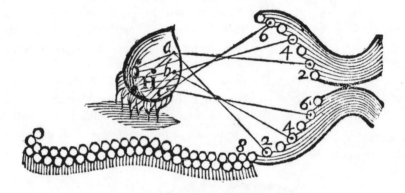

Figure 11: (iv) The pineal gland (close up) (*L'Homme*, AT 11, Fig. 1, 24, 27, 32).

(Fig. 12).But his figures simply depict the structures of the brain. If they convey a narrative, it is about the opening of the body to the eyes of the scientific spectator, not about the function of those structures.[20]

Comparing organs with machines was, as I have said, not at all new. What is new in Descartes is a determined, systematic attempt to analyze *all* the vegetative and sensitive functions of the body in the manner in which machines had long been analyzed. That program survived the demise of most of Descartes' particular hypotheses. It leaves its mark in lists, like Boerhaave's list quoted earlier, of mechanical devices occurring in the body, and in this striking passage from Giorgio Baglivi, a prominent iatromechanician:

---

[20] On anatomical drawings and their significance, see Sawday 1995, c. 5.

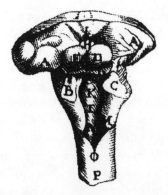

Figure 12: The pineal gland and environs (Bauhin *Theat. anat.* 1605, Liv. 3, Pl. 10, no. 10).

Examine attentively the physical economy of man: what do you find? The jaws, armed with teeth: what are these but pincers? The stomach is nothing but a retort; veins, arteries, the whole system of vessels, are hydraulic tubes; the heart is a spring; the viscera are nothing but filters and sieves; the lung is a bellows; and what are the muscles if not ropes? [. . .] all these phenomena must be referred to the laws of equilibrium, to those of the wedge, the rope, the spring, and the other elements of mechanics. (Baglivi *Praxis medica,* quoted in Guyénot 1941:158, Canguilhem 1975:104, Beaune 1980:224)

Sieves and hydraulic tubes fall only dubiously within the scope of the laws of equilibrium. Actual applications of those laws were few. But such objections bear more on the fate of the program than its character. The key was to isolate systems, and within them mechanisms, as simple as possible, whose operation, even if not described by quantitative relations, was intuitively evident in just the way that the operations of everyday tools are evident. As in the practice of building machines, those mechanisms would be portable, reusable, interchangeable.

Contrast this for a moment with the Aristotelian version of the analysis of capacities or *potentiæ*.[21] It begins with the objects and *actus* of a *potentia*. To understand vision, and to distinguish it from hearing or touching, one defines the visible. At that stage the analysis is largely independent of any description of the organ of sight. The organ and its qualities enter primarily in describing the *actus* of vision as one requiring certain dispositions in its matter: in vision, transparency and a certain degree of hardness; in hearing, undisturbed air within the organ.

The mechanical and the Aristotelian analysis converge in abstracting from matter. A gear can be made of metal or wood or any sufficiently hard yet workable stuff; the power of vision could be realized, for all we know, in transparent media other than the crystalline humor. But the character of the

---

[21] See Des Chene 2000, Chapter 6 for a discussion of Suárez and other authors on distinguishing the powers of the soul.

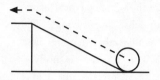

Figure 13: Inclined plane (AT 1:439).

abstraction differs. The gear does have a certain shape; its description consists in calculating from its shape (and those of the components it acts on and is acted on by) the transformation of forces it helps bring about. Nothing in the description corresponds to the object of vision, the visible formally considered. With respect to the intentions of its maker, it can be said to have an object of sorts, namely, motion of a sort to realize the aim with which the machine as a whole is built. But that object, Descartes will insist, exists only in the mind of the builder. Unlike the visible, a necessary element in the Aristotelian definition of vision (and therefore of the eye also), the motion that the gear is intended to make possible is no part of its definition. That the eye is a much more complicated machine than a gear makes no diiference: the visible has no role in its definition.

The principles by which mechanical components are classified likewise differ from those by which *potentiæ* are classified. Although the disposition of a simple machine will, when force is applied to it in the designated place, result in a particular motion, that does not determine its type. Instead the very names, 'inclined plane,' 'wedge,' 'wheel' are defined by shape; shape, first of all, and then the location of the force, load, and fulcrum, give rise to the three classes of lever. Simple machines are classified by their elements and the arrangement of those elements—their dispositions, in short.

The mechanical analysis of capacities may put one in mind of analysis and synthesis in the *Regulæ,* with the simple machines playing the role of the simple natures. But an ascent into the nimbus of method adds little, it seems to me, to an understanding of the theory of machines, or its application to physiology.[22] Mechanical analysis does open the possibility ultimately of explaining all their actions by reference to the modes of extension and the laws of motion. It has the important negative role of allowing natural philosophy to dispense with forms and faculties, or to claim that it has done so, and a positive role in forcing the philosopher to exercise his *ingenium* in devising mechanisms to replace them: the profusion of concepts noted earlier is one result. These do not, however, allow Descartes to arrive at ultimate explanations. But the promise of demonstrating effects by their causes (*Discours* 5, AT 6:45) contributes only a little to the persuasive force of the *Traité de l'Homme* and

---

[22] See, however, Garber 1993 for an illuminating account of the explanation of the rainbow in the *Météores*. According to Garber, the analysis of the refraction of light at the surface of a raindrop, by way of the prism (a kind of optical "simple machine") does reflect faithfully the procedures laid out in the *Regulæ.*

the *Description*. The greater part of that force is derived from the decomposition of the body-machine into mechanisms, and from comparisons of those mechanisms or parts of them to more or less well-understood, but in any case familiar, phenomena.

4.3 Instrument and Organ

Aristotle defines the soul as the "first *actus*" of an "organic body," a *corpus organicum*. The organs of the body, etymologically at least, are instruments of the soul.[23] To call something an instrument is to view it under a purpose; but it is also to assign to it a status among efficient causes. That is what interests me here.

In Aristotelian questions on the efficient cause, the instrumental cause, *causa instrumentalis*, is contrasted with the principal cause, *causa principalis;* the two together are sometimes called the total cause, *causa totalis*. The contrast turns up in questions on the efficacy of accidents in producing substances, on the efficacy of second causes generally, and on the "exemplary" cause, the paradigm of which is divine ideas in the creation of the world. Accidents are said to be the instruments of form (or of substance). In the generation of animals, for example, the heat of the seed is an instrument of the form of the seed. In law, the principals of a transaction are the actual purchasers, those who pay for and will own the item being bought; those who act on their behalf are usually not now called "instruments," though documents are sometimes so called. The core of the concept is that of *acting for*, but 'to the benefit of,' and 'in place of' or 'as the immediate agency of.'

In an exposition on principal and instrumental causes, Suárez records five ways in which the distinction, which he calls "very ambiguous and equivocal," was made.[24]

(i) The principal cause is that "to which the action is attributed properly and *simpliciter*" (without qualification) (Suárez *Disp.* 17§2n07, *Opera* 25:585). One difficulty with the criterion is that many things, like God, are not instrumental causes, and yet some actions whose total cause they are part of (as God is said to be by virtue of concurring in

---

[23] The Greek *organon* in Latin becomes *instrumentum*, though oddly enough *organicum* is usually left untranslated in the older Latin version of *De Anima*. On the related term *organisatio*, see Des Chene 2000, Chapter 4.3.

[24] For a brief discussion of Suárez's account, see Des Chene 1996:249 and Menn forthcoming.

the actions of second causes) are not attributed to them. That includes forms. Although Socrates' soul is "in a certain manner" a cause of his actions, we do not attribute Socrates' walking to his soul, but to Socrates himself. Suárez discounts the objection, on the grounds that, at least where the First Cause is concerned, the effects of second causes *are* properly attributable to him (586).

(ii) The principal cause is that which "by its own virtue inflows into the effect." 'Inflow' (*influere*) is what causes in general do: they "give" or "inflow" being (*esse*) into their effects (12§2no9, 587); in particular, the efficient cause initiates, sustains, or arrests change, the transient *esse* of *motus*. But heat, Suárez argues, warms objects by its own "proper and innate virtue," and yet is an instrumental, not a principal, cause. In other words, the instrument need not be a mere modulator of the action of the principal cause. It brings about, on behalf of the principal, an effect proper to itself.

(iii) The principal cause is that which "proximately and by its own influx inflows into the effect, or in the form that constitutes the effect" (12§2no10, 587). The instrumental cause, on the other hand, "does not touch upon the immediate effect or form, but on something previous, from which the form results." The instrumental cause is principal with respect to that "something previous," but instrumental with respect to the resulting form. For example,

> a saw is an instrument for making an artifact; the instrument [i.e., the saw] proximately effects only an incision, by impelling and causing local motion in intermediate parts, and with respect to that effect it is regarded as a principal cause: but from [that effect] an artificial form results, with respect to which it is regarded as an instrumental cause, because it does not attain to [the form] immediately by its own action; instead [the form] results from the prior effect.[25]

But that, Suárez says, is not quite right either. In particular, "in the instruments of art it sometimes seems that by the form of the instrument

---

[25] "Exemplis declaratur primo in artificialibus, nam serra, verbi gratia, est instrumentum ad tale artificium faciendum, quod instrumentum proxime solum efficit incisionem, impellendo, et loco movendo partes intermedias, et ad hunc effectum potius comparatur ut causa principalis: inde vero resultat forma artificialis, ad quam comparatur ut instrumentalis causa, quia illam non attingit immediate propria actione, sed solum resultat ex priori effectu" (Suárez *Disp.* 17§2no10, *Opera* 25:587).

the form of the effect is proximately attained, as when wax or money is impressed with a figure similar to that of the seal, although in fact figure is not made *per se,* but only location [Ubi] [. . . .] Therefore it may more generally be said that although instruments do not attain per se to figure or artificial form, that is not because they concur instrumentally [in the effect], but because that form is not otherwise producible" (17§2no11, 25:588). Elsewhere we are told that figure, alone among qualities, is never the *per se* effect of any cause, but only a *per accidens* effect. The *per se* effect of a saw is the moving of bits of wood, and the shape of the thing cut is only a *per accidens* result.[26]

(iv) The instrumental cause "acts only as it is moved by another," while the principal cause "has the power [*vim*] to operate by itself and without the motion of another" (17§2no13, 25:588).[27] Again that is not entirely adequate. 'To act as moved by another' can be understood in two ways.

First, a thing acts as moved by another if there is a real change brought about in it by the principal cause, prior to its own action. That may be true, Suárez says, for the instruments of art, but natural instruments do not need to be acted on by the principal cause beforehand: "The heat of fire is its instrument in producing fire, and vital heat, or the nutritive power, in producing flesh, but no one can think up any prior motion in them. Or again, the seed is called the instrument of the generator, but after it is separated [from the generator], it receives no new action from the principal agent" (17§2no14, 25:589).

Second, a thing acts as moved by another if it acts in that other's stead, as a subordinate or adjunct. Now every instrument does act in another's stead, but so too do all second causes. Yet they are not instruments of God, because they act "by their own intrinsic form," even if they do not act independently.

(v) The instrumental cause "is most properly said to be that which concurs or is elevated to the point of effecting something more noble than itself, or beyond its proper measure of perfection and action, like heat in-

---

[26] Figure, for the Aristotelian, is essentially all there is to artificial form. Natural things have no *potentia* to produce figure *per se*. Even though a sculptor intends the figure of a statue *per se*, "still the action, by which he executes his design, does not terminate *per se* and immediately in such a form, but in some other mode, from which such a form results" (Suárez *Disp.* 16§2no18, *Opera* 25:580). See Des Chene 1996:246–247.

[27] It is under this heading, perhaps, that we may put Descartes' reference to reason as a "universal instrument" in the *Discours* (AT 6:57). Reasoning, the combination of ideas for the purpose of judgment, occurs only at the behest of the will, and the formation of perceptual ideas only on the occasion of changes in the sense organs or the brain.

sofar as it concurs in the production of flesh" (17§2no16, 25:590).
Even in artificial things, "the form principally intended by the artificer
is not attained immediately and *per se* by the instrument, but results
from it; but however this occurs, [the form] exceeds in some way the
power of the instrument considered in itself, and is brought about by it
insofar as it is directed and moved by art."[28]

From Suárez's exposition one may gather, first of all, that 'instrument' is no
less complex a notion than 'machine.' The instrument is a subordinate cause,
a vice-cause, first of all; in many cases—and in all cases where the instrument is an artifact—it acts "as moved by another," both in the weak sense of
not acting independently and (if artificial) in the stronger sense of requiring
a prior motion from the principal cause; and, though Suárez does not regard this as counting for much, it acts in another's stead. The effect of the
principal cause is not brought about immediately by the instrument, but "results" (a word often used to denote indirect causal relations) from something
else that the instrument does bring about immediately. A saw simply pushes
bits of wood around; the shape of the thing it cuts results from that motion,
which alone is caused immediately and *per se* by the saw. Finally, the instrument is raised above its proper place in the order of things when it acts on behalf of a principal: from heat comes flesh, and from a printing press, metaphysical tracts.

The organs of the body are, according to Suárez's explication, instruments
of the soul, except perhaps in the sense of acting in its stead. (The seed, if it
is a kind of detached organ, does act in the place of the generator, which may
no longer be alive—consider the fate of the male praying mantis.) The proper
effect of the heart is to heat the blood; the nutrition and growth that result
from the movement of the blood are actions of the soul or its powers, which
are thus the principal causes of nutrition and growth. The animal, which,
though not generated by the heart, does depend on it for its continued existence, is in the Aristotelian way of thinking, more noble than the heart, because it alone is a full-fledged substance.

Artifacts, and machines in particular, are instruments *par excellence*. We often do attribute an action not to the instrument with which it is performed,
but to the principal cause, the agent who performs it. There are, of course,
degrees. It is less jarring to say that a computer did something than that a

---

[28] "In illis [sc. artificialibus], ut dixi, forma præcipue intenta ab artifice non attingatur per
se et immediate ab instrumento, sed resultet; quocunque tamen modo fiat, excedit aliquo modo
vim instrumenti secundum se considerati, et fit ab illo, prout ab arte dirigitur et movetur"
(Suárez *Disp.* 17§2no18, *Opera* 25:590).

toothpick did. But it is significant that agency is most naturally attributed to instruments when they go out of control: if a car had no driver or if its driver was incapacitated, then *it* crashed through the barrier, but if the driver was in full control, then the *driver* did. Moreover, most machines do require a prior action by the agent for their own actions, and as in the case of the saw, its immediate effect is something from which the action attributed to the agent results. The gear transforms force; from that transformation there results some useful effect—the provision of water for irrigation, for example. Even the fountains that Descartes dwells on in *L'Homme* were said to bring about pleasure in the spectator by making water move.

For Descartes, both the animal itself and its organs are machines. Take organs first: in what sense *are* they organs, that is, instruments? Certainly not in the last sense, the one Suárez prefers. Descartes will have no truck with degrees of nobility in nonhuman nature. The animal, the whole machine, is not more perfect than the mechanisms within it; nor is a machine or mechanism more perfect than a featureless lump of earth. Indeed the restriction in a mechanistic physiology to local efficient causes eliminates all but the fourth criterion[29]—except as applied to the human body when joined with the soul. Only in a body united with a soul are the organs of the body subordinated to anything. Only then can we distinguish readily between what the organ effects immediately and *per se,* and what results from that effect. The *aquæ fortes* in the stomach, as we have seen, begin to work on the stomach itself when the stomach is empty: their immediate effect is the dissolution of the parts of matter they come into contact with. In the union of body and soul, there is a further effect—the sensation of hunger—that results from the immediate effect of the *aquæ fortes* in the body. But in animals, there is just the pursuit of food.

As for the whole animal, to refer to the "machine of the body" is, first of all, to deny it proper agency. Like any Cartesian body, the animal is an agent only *quoad nos,* only with respect to a way of thinking that, though useful in everyday life, has no basis in the nature of things. The principle of life, as we have seen, is the heat of the heart. Heat—including the *feu sans lumière*—is rapid motion, caused by prior motion of the same sort. 'Caused by' simply means: resulting from collisions. In a collision, each of the colliding bodies has as strong—or as weak—a claim to the title of agent as the other: but since Cartesian matter, being no more than extension, is utterly inert, in fact no body is an agent. Genuine agency pertains to spiritual substances alone.

Perhaps the only way for Descartes to distinguish organs from the whole

---

[29] In a world where *everything* is moved by another except God, the only principal cause is God. All corporeal things are, by the fourth criterion, instruments.

animal in the causal order is to regard the machine of the *whole* body as the only entity of which one may say that it moves itself, while the organs do not. That will be true in the limited sense of 'self-mover' introduced earlier in §1.2. The body is capable of restoring, through its cycles of motion, the initial conditions under which certain motions—digestion, respiration, the pulse—are bound to occur. In particular, it brings into itself the fuel consumed by the *feu sans lumière*. No organ alone does so. The heart, lungs, stomach and so forth can in that sense be said to be subordinate to the whole body in their operation.

But are the organs of the body truly the instruments of the soul? The ear is part of a causal chain that yields sensations of sound in the soul. It does not have sensations as its immediate effect—its only immediate effect is to move threads in the auditory nerves. Indeed the logic of instrumentality, if I may call it that, would have it that if the organ of hearing is truly an *instrument,* then the principal cause of hearing must be either

(i) the vibrations in the air that move the organ of hearing, or
(ii) the soul in which the principal effect—the sensation—occurs, or
(iii) something else that causes that sensation, either immediately in the mind or else by way of the vibration.

The first choice is an unhappy one for the reasons just mentioned. If bodies lack genuine agency even with respect to their interactions with one another, then *a fortiori* they lack agency with respect to the soul. The second runs contrary to Descartes' consistent characterization of sensations as "passions," that is, as changes undergone by the soul. The argument of the Sixth *Meditation* requires that the soul *not* be the cause of its own sensations. As for the third, the obvious candidate for the principal cause is God. If he causes the sensation immediately in the soul, then the vibrations in the air (and the movements in the body caused by them) would be at best occasional causes of the sensation, as in Malebranche or Geulincx. God must therefore be the principal cause of the sensation *by way of* the vibrations in the air, the movements in the organ of hearing, and the subsequent movements in the nerves and brain. But then the organ of hearing is God's instrument, not the soul's.

In Aristotelian physiology and in the medical literature of Descartes' time, the body is subordinated to the soul (as in general matter is to form), and the organs of the body are arranged into an order of perfection. The organism as a whole is a kind of polity within which some parts by nature serve others, and are their instruments. In Descartes' physiology, on the other hand, though some parts of the body may be needed more for its continued operation than others, there is no conceptual basis by which to justify any hierarchy among them. In the human case, moreover, since the body is capable, in principle, of functioning autonomously, its parts cannot be said to be instruments of the soul by virtue of the mere fact that the body lives. Instead there are two reasons one might give for calling the parts of the body, or all of it, instruments of the soul. The first and more obvious reason is that the body is an instrument of the soul in at least the fourth and fifth of the senses canvassed by Suárez when by an act of will we make our body move in order to accomplish some end. The body is indeed moved by another—the soul. Its action may well bring about *per accidens* an effect higher than any that it alone can cause—your thoughts as you read these words, for example. The second reason is that the body functions to maintain not only its own continued operations but the union of body and soul, for example, by supplying through the senses the practical guidance that the soul needs to preserve the union. Here, however, the principal cause must be God. The body is God's instrument, not the soul's.

4.4 Artifacts

The paradigmatic machine is an artifact, a human invention. God, creator not only of the matter of all things, but of their forms, is the *artifex maximus:*

> On Descartes' page is Nature's self disclosed,
> From whose ingenious mind all men are born.
> Animals, that moved themselves, are moved by art;
> The copious machine betrays its maker, God.[30]

An ancient trope, certainly, and a mainstay of apologetics. But not without its perils for the Christian Aristotelian. The products of *human* art (in the broad sense: *ars* or *tekhné*, anything designed and made by humans) have *per se* no substantial form: a clock is not a substance but an assemblage of substances.[31]

---

[30] Bartholin *Carmina* 217, quoted in AT 5:570:
Cartesii chartis natura recluditur ipsa,
Cujus ab ingenio nascitur omnis homo.
Namque animal quod sponte prius, nunc arte movetur,
Prodit & artificem machina plena Deum.

[31] A minor—but in looking forward a significant—exception was made for the substances produced by alchemists. If an alchemist did manage to produce gold from some other substance, that gold would be identical in kind with the gold made by God, and would therefore be endowed with all its natural powers.

Even a single gear is not a substance by virtue of being a gear, but by virtue of being a lump of metal. Its metallic form alone determines its nature; the shape applied to it by the watchmaker is quite incidental to its natural powers and dispositions. The artifact has for that reason no agency of itself. If it acts— if it falls to the ground, say—it does so by virtue of its proximate matter, not its artifactual form.

Nevertheless the trope would have it that all created things are somehow artifacts. Human artifacts have no agency: how is it that divine artifacts do? Humans and God have the *habitus poeticus sive mechanicus,* as Goclenius calls it (Goclenius *Lexicon* s.v. 'ars'): how is it that the divine *habitus* differs so much from ours as to produce things having genuine form and active powers? To call something an artifact is to regard it as (i) made, not eternal; (ii) having a "design" which was intended as such by its maker in the making of it. So far nothing precludes an art of genuine form. Yet the Aristotelian denies genuine form to artifacts.

The answer lies in the order of being, and its correspondence with the causal order. The more noble, the more perfect, can produce the less noble, the less perfect, but not conversely. But there is a further fillip to the doctrine when it comes to the generation of animals. The animal soul, and *a fortiori* the human soul, cannot be produced even by its like. Only the celestial intelligences and God are capable of making an animal soul, and only God can make a human soul. In either case only the more noble cause has enough power to produce something with active, autonomous powers of its own. An *excess* is required when we reach the level of sentient existence. The *same* will not do. Indeed among Aristotelians it was a serious question whether on the part of the parents there is any action "that intrinsically attains to the rational soul, uniting it with matter," or whether the role of the parents (and the seed) is limited to "preparing the matter in which, when it is perfectly disposed, God will infuse the soul and unite it with the body" (Coimbra *In de Gen.* IC4qI3aI, pII4).

The tension within Christian philosophy between the conception of God as the divine artist or Demiurge and the attribution to created things of genuine agency is resolved by the correspondence of the order of perfection with that of causal efficacy, orders that cannot be flouted. The distinction between animal souls and inanimate material forms is not one of degree—i.e., not one of increased number or fineness of parts, or intenser qualities—but of perfection. The material substrate of the animal form, as we have seen, provides a necessary basis for its operations, but however ingenious it may be, it does not itself constitute the form of the animal. Hence it is possible for the animal soul to have powers not found in inanimate things; but, correspondingly, it is impossible for the animal soul to introduce its like into new matter without assistance from yet more perfect forms. Once that is established, the apologist is free to expound upon the ingenuity and complexity of the organization of the animal:

> Nor is it to be judged that the fabrication of living bodies lies beyond the power of nature. Rather it is equally admirable that the infinite power of the divine artifex should bestow faculties upon material, concrete nature for its perfecting. [Nor it is to be judged beyond the power of nature] that the seed should observe a correct and constant order in its work, and obey the sequence of generation, and arrive at a certain end; the reason or counsel it needs is supplied sufficiently by certain numbers and measures, so to speak, by which it is directed by the author of nature, from whom for that purpose it accepts a propensity and virtue having, in some way, the power of art. (Coimbra *In de Gen.* 1c4q24a1, p190)

The organizing of the *body,* as opposed to the introduction of the *soul,* is, in the view of the Coimbrans, quite within the powers of the seed. There is here a division of labor: to nature the works of which nature is capable, to God the works that exceed nature.

The analogy between human and divine artifacts, then, does not preclude the attribution of souls to animals and plants. But in the Coimbrans' division of labor there is a significant difference from those theories that attributed *organisatio* also to supramundane powers. The *artifice* of the body is the work of nature: in that sense the *bête-machine* exists long before Descartes. More humble than the making of souls, that work is accomplished by God at second hand, through the seed.

What remains for Descartes, then, is not to argue that the animal is a machine but that it is *only* a machine. Hence the strategy of *L'Homme* and the *Description:* to show that "the disposition of the organs is by itself sufficient to produce in us all the movements that are not at all determined by our thought," so that we will have no more occasion to believe that a *soul* produces those movements than "to judge that there is a soul in a clock, because it shows the hours" (*Descrip.* AT 11:226).[32] Such an argument cannot attain to certainty. It is not impossible that a soul could produce those movements, as the Aristotelians hold. It cannot both be true that the machine has a soul

---

[32] Hence although it is correct to observe, as Baker and Morris do (see Chapter 1, n. 1), that machines were among the marvels of the early modern period, it is also true that to call living things machines was indeed to diminish them, since to do so meant denying that they have souls, and the animate is intrinsically more perfect than the inanimate.

and that it does not (though one might certainly consider dividing the cases: animals and plants, or lower and higher animals). But there is, it would seem, nothing that would force the issue. Even the sufficiency of mechanical operations in simulating the phenomena of life need not compel us to abandon a benign probabilism.

But the oldest game of philosophers is to force the issue.[33] In Descartes' hands, the trope of the machine is turned to that end, to make it *seem* as if 'mechanical' and 'animate' were antagonistic, so that the Aristotelian soul is ruled out by the existence of mechanical processes.[34] For the Aristotelian, that is not at all true. Each kind of cause has its place in the explanation of the phenomena (whose descriptions to some extent themselves vary with the expected kind of explanation). Material causes (which could as well be mechanical) effect the disposition of the matter of the fetus; the soul or its minister in the seed coordinates those causes; a higher form creates the soul. It need not be that the organizing is distinct from any of the material goings on, nor that a fortuitous combination of matters could not do the same (as in spontaneous generation). That possibility in no way precludes the operation of higher causes in the more usual instances of generation.

The exclusiveness of mechanical causes cannot be argued outright. Descartes' aim is rather to undermine alternatives, both by exposing the history

---

[33] In *Les maîtres de vérité de la Grèce archaïque*, Marcel Detienne argues that dialectic originated in a rivalry between the poets, who by virtue of the authority vested in them by memory (Mnemosyne, a divine power) were granted "access to the beyond," the suprasensible, from which they produced a discourse "marked by ambiguity," and the Sophists and philosophers, who "opened a new intellectual regime, that of argument, of the principle of non-contradiction" and of inquiry into the meanings of words (Detienne 1967/1994:6–7; see also 177 on the division between philosophers and Sophists). In the seventeenth century, as many authors have noted, the system of *authority*, in which validity was conferred on *experientia*, the data of science, by reference to everyday experience and to established texts, broke down and was replaced by new systems of validation. The exhibition of the machine, by which the reader of a work like *L'Homme* could be persuaded first of the *possibility* of mechanical explanation and second of the *actuality* of the particular mechanisms displayed, was one element of the new system. To this was added the arguments by which the issue was forced between mechanical explanation and explanation by forms and qualities: *either* the animal is a machine *or* it has a soul; what we now call emergent properties are, in Descartes' version of mechanism, effectively ruled out.

[34] One may see here an application of what Stephen Yablo has called the "causal exclusion principle" (Yablo 1997). If we can find a sufficient cause for the raising of my arm in the physical processes of my brain, then any other cause (an act of will, say) is superfluous. The version of the argument in Descartes' physiology is (bearing in mind that for the Aristotelian the animal soul, though corporeal, is nobler than the forms of its constituents) what Yablo calls the "argument from below," which amounts to this: "with each physical effect causally guaranteed by its physical antecedents, what is there left for its mental antecedents to do?" (253). The argument from below was often employed against substantial forms in the seventeenth century. Its use against every version of dualism first occurs, if Yablo is right, in the twentieth.

of our prejudice in their favor, and by making it seem as if anyone who agrees with the mechanical explanations in his physiology must *thereby* agree also that no other cause is present (and not merely that no other cause *need be* present to produce the phenomena as Descartes describes them). Consider the presentation of the machine to sight and imagination. A picture cannot easily convey a negation (what should a picture of Caesar's *not* crossing the Firth of Forth look like?). But it can do everything in its power to insinuate that what it shows is all there is to show.[35] The primary means of doing that are detail—the more circumstantial the better—, and clarity and distinctness, whether in imagining or in graphic technique; subject to a condition of exhibiting only what is relevant to the explanation at hand. The implication of the picture is that you, the observer, are seeing *all that matters,* and seeing it *as it really is.*

In that respect, the machines illustrated in books like Salomon de Caus's have, so far as the implication of the graphic is concerned, no secrets. Their depiction differs greatly in that respect from the illustrations in a work like Robert Fludd's *Utriusque cosmi historiam,* some of which seem to conceal (not without suggesting that revelation is to be had elsewhere) as much as they reveal.[36] That is not to deny that mechanical philosophers too had their trade secrets, or that features essential to the actual operation of a machine (like grease and leather gaskets) might not be absent from the illustration—features one might suppose remained within the tradition of craft knowledge that books of machines were making public and *raisonnée.* But the ethic, if I may call it that, of mechanical illustration is to show everything the reader needs to know (supplemented, if need be, by shop-lore) in order to understand and build the machine depicted. What you wish to keep secret, you put in a black box, you do not shroud it in riddles and allegory.[37]

The machine, then, is open, without secrets, something that can be artic-

---

[35] In a full-blown theory of pictorial representation, one would need first to characterize what a picture is intended to show, and only then proceed to the means by which a picture may purport to be an exhaustive depiction. The first figure in *L'Homme* [see Fig. 11(i)] is not intended to show the whole body, nor even all the parts of the organs it does depict. Rather it may be taken to purport to show all that an observer could see *at that scale* given normal visual acuity.

[36] For examples from Robert Fludd, see Huffman 1988, Figures 6 (*Utriusque cosmi,* 1617–1626), 11, and 12 (*Medicina catholica,* 1629–1631). Caus, at the end of the Third Book of his work, mentions "certain very singular machines, which I take to be secret, and among them one that will present a music more perfect than any human being can make," but only in the course of promising another book (Caus *Forces mouvantes* Liv. 3, f8 (after Prob. 17).

[37] The openness of the laboratory of virtuosi like Boyle is sometimes contrasted with the occult chambers of the alchemist. But that is not the whole story: with new forms of publicity (the report, for example, published in the transactions of a learned organization) new forms of secrecy are also invented.

ulated completely in an *ekphrasis*.[38] That is in keeping with its other aspects, of being artifact and instrument. The artifact owes its artifactual form entirely to the intention of its maker. Having an accurate, complete picture of its design is (in theory if not in fact, since making almost any machine requires tinkering) sufficient for its fabrication. The blueprint, successor to the pen and pencil of early modern engineers, is the type itself of a fully explicit intention.

That the intentions behind animal-machines lie in the mind of God, and may never be known to us, does not tell against calling them machines. It is true that in the *Discours* Descartes presents his physiology within the scope of an epistemology of clear and distinct ideas, and demonstrative knowledge based on them. It fails to be "in the same style as the rest," he says, only because he doesn't know enough yet (*Discours* 5, AT 6:45). The ontology of the machine is indeed one of extension and its modes, and in principle the operations of machines are subject to geometric demonstration. Nevertheless, with the partial exception of the optics of the eye, no explanation in *L'Homme,* the *Description,* or any of his other writings on physiology is anything but *a posteriori,* anything but hypothesis, accompanied by hints at an argument for uniqueness.[39] The epistemological allure of the machine, it seems to me, ought to be traced not so much to explicit doctrines of method as to the intuitive appeal of the blueprint, an appeal that requires only that the design should *exist*—be it in the mind of God—and not that it should be entirely knowable to us.

Descartes' explanation of the formation of the sense organs is a case in point (§2.2). Another is the explanation of the passions in *L'Homme.* The configuration of the animal spirits can vary along four dimensions: abundance, size, agitation, and uniformity in size. It is "by means of these four differences that all the various humors or natural inclinations in us (at least insofar as they do not depend on the constitution of the brain, or particular affections of the soul) are represented in the machine. If they are more abundant, they "excite

---

[38] *Ekphrasis* or description was an ancient literary genre that consisted in the detailed narrative description of a real or imagined painting. See Baxandall 1985:2–3 for an example from Libanius.

[39] In *L'Homme,* for example: "Finally, as for the rest of the things I have supposed, which cannot be grasped by any of the senses, they are all so simple and common, and even so few in number, that if you compare them with the diverse composition and marvelous artifice that appear in the structure of visible organs, you will have more grounds for thinking that I have left out some than that I have supposed any that do not exist. And, knowing that Nature acts always by the means that are the easiest of all and the simplest, you will perhaps not judge it possible to find any more similar to those she uses than those here proposed" (AT 11:200). Descartes *could* do better than this when faced with a genuine rival. Witness the reasons he gives for preferring his theory of the motion of the heart to Harvey's. One can only wish that he had accorded more opponents the same degree of respect.

movements entirely similar to those that in us are testified to by [the passions of] *benevolence, liberality,* and *love*"; if they are big and strong they cause movements like those that give rise to *confidence* and *boldness;* if uniform in figure and force, then *constancy;* if agitated, *promptness* (in wit), *diligence, desire;* if uniform in movement, then *tranquility of mind* (*L'Homme,* AT 11:166–167). What else is there to this but a transcription into mechanical dimensions of the phenomenological qualities of the passions? But at least we have a starting point, a schema to work from.[40]

When, in the early eighteenth century, the mechanical philosophy began to be tinged with pessimism, a pessimism we find expressed in the antimechanism of Stahl,[41] in the hesitations of Locke, or in Claude Perrault, who writes that "in the functions of animals there is something that cannot be explained even by all we know about the properties of corporeal things,"[42] it was in part because the allure had faded. So much so that in 1724 the *Journal des savants* could register only boredom at the appearance of yet another *Exercitatio physico-medica:* by then, after all, "an infinity of authors [had] written on the matter."[43] Naturally there were other reasons, including the failure of mechanistic theories to explain generation. But it seems to me that, as later with electricity and the computer, the rapid comprehensibility of comparison with the machine, which at first eases its path through regions of knowledge, begins eventually to tell against it. The most accessible new lands are colonized; what remains are highlands and wastes that can only bring out its limitations.

To conclude, I note one unsuspected consequence of the success of mechanism in the seventeenth century. For the Aristotelian, as I have said, the difference between the divine artificer and his human imitators is not one of degree, but of perfection. Only God and the celestial powers are active enough to make not only movers but *self*-movers. In the apologetics of mechanism, on the other hand, the difference becomes quantitative alone; that is in fact a selling point. God's machines have more parts, tinier parts, more intricate

---

[40] Bitbol-Hespériès notes that the four dimensions of the classification correspond to the four Hippocratic humors, themselves defined in terms of the four qualities, hot, cold, wet, and dry (*Le Monde/L'Homme* 194n139–195n140). It should be noted, however, that Descartes does attempt to draw consequences from the classification that would lend independent support to the properties attributed to the spirits in the various instances—for example, that anger makes the heart beat faster. The whole is best thought of on the analogy of the "rationalization" of myth: the Flood, say, becomes scientifically acceptable if it can be given a *prima facie* plausible account. Here traditional taxonomies of the passions are being saved for science.

[41] See Rothschuh 1968:153–154 (a passage from Stahl *Diss. inaug.* §20–24 in German translation); Hall 1969, c25; Duchesneau 1982:xix, 6–11; P. Hoffman 1982:205–210.

[42] Roger 1971:220, quoting Perrault *Méch. des anim.* 270–271.

[43] Roger 1971:207, quoting *Journal des Savants* (Nov 1724): 703.

parts; but there is no difference in nature or in our understanding between his machines and ours. The gap between his power of creation and ours of making has thereby been reduced. The remedy, foreshadowed in the *Meditations* by Descartes' arguments for the priority of the infinite over the finite, was to insist all the more strongly on the quantitative incommensurability of divine creation and human industry. Thus we find Leibniz proposing an actual infinity of machines within the animal. The power of the Creator was thereby kept beyond our grasp.

[5]

Jeux d'artifice

No reader of Descartes' early *Cogitationes privatæ* can fail to be struck by their fascination with illusion, concealment, and revelation. A series of optical illusions, for example:

> *Item,* to shape the tops of fences so that from a certain perspective they represent certain figures:
>
> *Item,* in a room, to make the rays of the sun, passing through certain openings, represent various numbers or figures:
>
> *Item,* to make tongues of fire, chariots of fire, and other figures appear in the air in a room; all by certain mirrors that gather rays into those points. (AT 10:215–216)

Optical illusions reappear in the *Météores,* when, having explained the rainbow, Descartes describes "an invention for making signs appear in the sky, which will cause great admiration in those who do not know their explanation" (*Météores* 8, AT 6:343–344). The *Dioptrique* devotes some pages to the *veridical* illusion by which the telescope makes faraway objects look near. Or again, mechanical illusions: in the *Cogitationes* we find the dancing statue mentioned at the beginning of this chapter. A few years later, the *Regulæ* mention, in a list of puzzles, a "basin [. . .] in the middle of which there stands a column, on which is set the figure of Tantalus making gestures as if to drink," but frustrated each time he does so (AT 10:436). The *Traité de l'Homme,* of course, contains a lengthy description of statues of mythical figures, that, powered by water, appear to flee one's approach (AT 11:120, 131–132).[1]

Artifice in service to illusion appears frequently enough in Descartes' works to be counted among his major preoccupations. That is no surprise, in view of their long association in Western philosophy, but in Descartes it is the illusions produced by machines, and not the more notorious illusions of visual art, that are foremost.[2] Artifice and illusion are conjoined in Descartes' scientific works with a third notion, that of simulation. The resulting complex figures in his thinking on method, on the ontology of natural philosophy, and on our æsthetic relation to nature.

A theme so prominent, and so central to the project of the *Meditations,* has not lacked commentators.[3] In this chapter my aim is confined to drawing out some connections between simulation, artifice, and illusion, with special regard to machines, and to introduce the question of the role of intentions— especially those characteristic of the use of machines to simulate or produce illusions—in the individuation of machines. The question of intentions will return in §6.4.

5.1 Simulation, Illusion, and Questions of Method

Simulation, insofar as it amounts to the explanation of a range of phenomena by systematically reproducing them in a model, may be put under the heading of "enumeration" in the *Regulæ,* or of those hypotheses that, like the decipherment of a coded letter, can attain to moral certainty only (*PP* 4§204–205, AT 8/1:327–328). If the imagined man-machine can be shown to operate as we do, then provided that we have explained all the phenomena by its

---

[1] A distinction should be made here between illusions in which we are knowing, willing collaborators, and illusions that conceal their nature as such. Perspective drawing produces an illusion of depth. But except in certain uses of *trompe-l'œil* painting, it does not pretend to be other than drawing—markings on a two-dimensional surface. Tantalus and the other hydraulic statues mentioned by Descartes are illusions of the first kind. It would be interesting to know if, for example, any of the bird-song machines so vaunted for their realism was ever actually used to fool someone.

[2] Passages like those from the *Cogitationes* never, to my knowledge, mention the beauty of the illusions produced. At most such apparitions produce admiration, but only in the ignorant. Cavaillé observes that the "science des miracles"—Descartes' name for the part of mathematics devoted to production of optical illusions—is a "simulation dénonciatrice" that shows the vanity of magic (see To **** (probably Mersenne), Sep 1629?, AT 1:21, and Cavaillé 1991:48). One well-known textbook in the science of miracles was Niceron's *Perspectives* (1638).

[3] Among the more recent longer studies, see Romanowski 1974, Grimaldi 1978, c.4, and Cavaillé's excellent *La fable du monde* (1991).

means, and those phenomena are sufficiently plentiful and varied, we can infer that the human body—if it is a machine at all—is a machine of the sort we have imagined. The inference may be strengthened by an appeal to simplicity, most effectively when sifting through hypotheses of the same basic type: hypotheses that all take the mechanical philosophy for granted, for example. But when Cartesian and Aristotelian or Galenist hypotheses are being compared, the application of the criterion is unclear and uncertain. How should we weigh trade-offs between faculties and mechanisms, or between qualities and configurations? Is an ad hoc pulsific power more or less simple than an equally ad hoc supposition about the shapes of blood particles?[4] That is why I inserted the qualification 'if it is a machine at all': As we have seen, to demonstrate the machine-hypothesis in broadest terms is quite another matter. For that purpose, any *sufficient* simulation will do, first of all—it needn't be the simplest. The further task is to show that faculties, forms, and the rest, even though, *pro tempore,* they may be sufficient too, are ruled out from the beginning, *before* criteria of simplicity are applied.

However one might settle such questions, for Descartes the virtue of simulation by machine was that it did not stray beyond the domain of entities admitted in his physics—and, I think, that it *provoked* suppositions. A mind well-stocked with mechanisms like those used and reused in the machines of Caus and Ramelli will find congenial the task of finding a mechanism that would behave in a specified manner. We find Descartes writing an unusually detailed description for Mersenne of an experimental setup for determining whether a cannonball shot straight up will fall back to earth (To Mersenne 15 May 1634, AT 1:293), or, in the *Dioptrique,* giving plans for the construction of a lens-grinding device (*Diop.* 10, AT 6:216–224). Other letters find him advising Ferrier and Huygens on practical details of the craft.

But those same instances indicate the weakness of simulation. The engineer may devise mechanisms *à sa fantaisie,* so long as they do the job, just as the philosopher telling fables may feign matter to his liking. Naturally one would like to economize on materials and labor, but often a wide variety of designs will accomplish the same task. The natural philosopher who is *not* telling fables cannot allow imagination to go where it will. Though Descartes occasionally writes as if empirical equivalence were enough (*PP* 4§204, AT 8/1:327),[5] his behavior in controversy and the promise provided by divine

---

[4] On ad hoc mechanical hypotheses, see Gabbey 1990.

[5] Though Descartes says that he will consider himself satisfied to have written things that "accurately answer to all the phenomena of nature," this is immediately put in the setting of the uses of everyday life and of the practical arts of Medicine and Mechanics, which have "only those things that are accessible to the senses [*sensilia*], and numbered among the phenomena of

veracity that we should be able to discover many truths, or even all the truths of nature (*Med.* 6, AT 7:80), argues decisively in favor of his being what we would call a scientific realist. Alone, simulation is not enough, if Nature, which we may suppose uses the simplest means, is too clever for us.

To simulate something is make something else that operates as it does or would.[6] Two clocks, precisely similar on the outside, will simulate each other, even if "within they are constituted by very dissimilar arrangements of gears" (*PP* 4§204, AT 8/1:327). There is no question here of illusion, still less of deception. Illusion arises when the gears are hidden from us. Bemused by the phenomena, the vigilance of reason perhaps lulled by admiration, we may well accept whatever explanation of them is suggested to us by the clockmaker, or by our *præjudicia,* the preconceptions we adopt when the soul is still in thrall to the sensations and passions that come to it by way of its union with the body.

The threat of *systematic* illusion, imposed on us by a deceiving God or *malin génie,* requires much stronger measures to be defeated than mere experiments or judgments of simplicity. Or rather, that experiments and judgments of simplicity should have any role at all in the search for truth requires that we should have already defeated the *malin génie,* by the ruthless but efficient means of showing that he cannot exist. It is not my business here to delve into the intricacies of hyperbolic doubt, but only to point out that behind the statues of *L'Homme* a more sinister figure lurks: a God or demon who by substituting his own power for that of ordinary second causes can falsify any hypothesis whatever. In Suárez the thought is entertained that all the vital processes of living things, or indeed any effect of second causes, can be brought about immediately by God. If that possibility is taken seriously, and entered in the books alongside less "metaphysical" candidates, then the science of life, with all science, is in jeopardy. That is the dark side of the divine Artificer.

5.2 The Double Twist

Overcoming hyperbolic doubt requires entertaining radical hypotheses about the nature of the world and the things in it. The result of overcoming that doubt is an unshakable assurance first that I, and then that God exists, which

---

nature, for their end." (*PP* 4§204, AT 8/1:327). Truth is another matter, as the next two paragraphs of §204 demonstrate.

[6] Resemblance in operation is essential to simulation; resemblance in outward appearance or inward structure is not. I will return to this in §5.3.

puts to rest any *præjudicia* to the contrary. Other, less fundamental, *præjudicia* likewise require the medicine of simulation & deception. In the theory of the animal-machine, however, there is a dizzying, if not perplexing, double twist.

In the *Traité de l'Homme,* we are presented with, as part of the fable, statues made by God to resemble us as much as possible. No deception is intimated, because the world of *Le Monde* exists in what Descartes ironically refers to as "imaginary space," so far from ours that ours cannot be seen from it—nor presumably can it be seen from ours. Even if, as Descartes optimistically proposes to Mersenne, lenses can be ground that would enable us to learn if the Moon is inhabited, they will not show us the statues of that distant world. But in the *Discours,* we are brought face to face with machines, "incomparably better ordered" and "having movements more admirable than any that can be invented by men," and that bear, moreover, "the figure of an ape, or some other animal without reason"; then we encounter others of similar make that "have the semblance of our bodies." The animal-machines will succeed in deceiving us, because "we would have no way to recognize that they were not in every way of the same nature" as the animals we are familiar with. After all, the animals we are familiar with *are* machines of incomparable quality "made by the hands of God." It is like trying to tell water from water. God cannot deceive us by placing animal-machines before us, unless he were trying to imitate the character in the joke told by Freud who deceives his friend by telling him he is going to Cracow when in fact he is going to Cracow. The human-machines, on the other hand, might succeed in deceiving us, were it not that we have "two very certain ways to recognize that they are not, for all that, genuine men." One is the use of language, the other the use of reason. These will protect us against a certain kind of illusion or deception which would be otherwise inevitable, since the machines may resemble us in every aspect perceptible to sense.

The animal-machine cannot be used to deceive us, except in the less important respect of introducing errors of detail into our physiology. But the man-machine can deceive us. It can produce—perhaps for far longer than we have time to uncover the deception—the words and deeds of a being endowed with reason. Only by reflecting on the quality of reason as a "universal instrument," and contrasting it with the necessarily finite repertoire of "particular dispositions" realizable in a machine, do we have the chance to unmask the machine, even given infinite time.[7] When Descartes writes that it is "morally impossible that there should be enough different [dispositions] in a machine

---

[7] Baker & Morris 1996:88–89. Note that the universality of reason, like the infiniteness of God, is not merely an indefinite extension of the finite capacities of machines.

to make it act in all occurrences of life in the same manner that our reason causes us to act" (*Discours* 5, AT 6:56–57), the *all* must be taken seriously.

That is the first twist. The animal-machine, though it be presented as a fable, cannot deceive us; nor indeed can the man-machine, insofar as those actions which occur in the body independently of our will are concerned. When God makes a body that resembles ours as much as possible, he makes a human body: what could prevent him from doing so?

The second twist occurs in a letter to Reneri for Pollot, a fellow soldier of Descartes' who had sent one of the few letters Descartes received in response to the *Discours*. Pollot argues that

> experience shows that animals make their affections and passions understood by their sort of language, and by a number of signs they show their anger, fear, love, pain, their regret at having done wrong [. . .] it is evident that animals perform their operations by a principle more excellent than the necessity originating from the disposition of their organs; namely by an instinct, which will never be found in a machine, which has neither passion nor affection as animals do. (Pollot to Reneri for Descartes, Feb 1638, AT 1:512)

This was a standard riposte to Galenist and atomist dismissals of animal souls. The mention of instinct is perhaps rather Stoic than Aristotelian, but it is an objection Descartes might well have heard in his student days.

A polite but uninformative reply from Descartes almost immediately followed, in which Descartes says he will set aside a copy of the *Géométrie* for Pollot, at that moment a prisoner in the fort of Callo (AT 2:673, note a; To Pollot, 12 Feb 1638; AT 1:518). A month later Descartes, in a letter never received by Pollot, answered all fifteen of his queries. To the query about animals he writes:

> It is certain that the resemblance between the greater part of the actions of animals and our own actions has given us, from the beginning of our life, so many occasions to judge that they act by an internal principle similar to that which is in us, that is, by means of a soul which has feelings and passions like ours, that we are naturally prepossessed with this opinion.

To deny it therefore exposes one to the ridicule of "children and weak minds." But those who desire truth must "shed the opinions by which they have been forestalled since childhood." Then a thought-experiment:

> consider what judgment a man would make, who had been raised all his life in a place where no animals but men had ever been seen, and where, being quite

devoted to the study of Mechanics, he had fabricated or helped to fabricate a number of automata, of which some had the figure of a man, others that of a horse, others that of a dog, a bird, etc., and which walked, ate, and breathed— in short, imitated as much as possible all the other actions of the animals they resembled, without omitting even the signs we use to bear witness to our passions.

He then refers to the two ways of distinguishing people from machines that look like people, and continues:

> One must, I say, consider what judgment this man would make of the animals among us, when he saw them; above all if he were imbued with the knowledge of God, or at least if he had observed how inferior all the effort [*industrie*] that men put into their works is to that which nature has put into the composition of plants [. . .] so that he believed firmly that if God or nature had formed certain automata that imitated our actions, they would imitate them more perfectly and would be incomparably more industriously made than any of those that can be invented by men. Now there is no doubt that this man, seeing the animals among us, and observing in their actions the same two things that make those actions different from ours as he had noticed in automata, would not judge that there was in them any true sensation nor any true passion as in us, but only that these were automata which, being composed by nature, would be incomparably more accomplished than the automata he himself had made hitherto. (AT 2:40−41)[8]

Finally, one must consider whether that judgment, made in full knowledge of the reasons for it, is less credible than the judgment "we made when we were children," a judgment now adhered to solely out of habit, and founded only on the resemblance between some of the actions of animals and some of ours. The answer, of course, is that it isn't.

No doubt Pollot, had he been able to receive this letter, would have judged it well worth the stamp. But the argument is peculiar. It sets against us a man

---

[8] Descartes responds to a query of Mersenne with a brief reprise of the thought-experiment: "if we were as accustomed to see automata which imitated perfectly all of our actions that they can imitate, and to take them only for automata, we would not at all doubt that all reasonless animals are also automata, because we would find that they differ from us in all the same things" (AT 3:121). In this version, it is crucial that we are *not* deceived by the automata (which in the earlier version was assured by having *only* automata in the first scene, and by having the stranger build them himself). They form a class *indistinguishable* from that of animals, since they imitate their actions perfectly, and yet by some prior knowledge the members of the class are known to us as such. That perfect imitation is possible Descartes confirms in referring Mersenne to the *Traité de l'Homme*.

who has had *no experience* of animals and yet is in a better position, after a little while, to judge their nature than we are. A picture of expertise only too familiar now.

The animals of our prepossessed thought, which have affections and passions and souls, turn out to be illusory, and the animals in the conception of the stranger, who has lived among what he knows full well are machines, are the animals that have been beside us all along. That is the second twist. All unknowing, the stranger was building crude but genuine *animals* (crude but genuine as the hesitant and laborious ABC's of a first-grader are crude but genuine, that is really A's, B's, and C's). Joining us, he now recognizes their incomparably more accomplished cousins. ("Incomparably," one might ask, how? If there is not, after all, to be a difference in kind, it must be only that their construction is not humanly *feasible,* that it is a sheer matter of the size of our fingers and the capacities of our intellects, unable perhaps to form a clear and distinct idea of their whole design. Feasibility, in the Cartesian world, is of no importance except where practical interests are at stake.) [9] What matters in the search for truth is the existence of a blueprint, which the stranger infers with assurance from the resemblance of animals to the automata fabricated by him, and their lack of resemblance to humans in the crucial respects.

It is not resemblance alone that creates illusion. We do make an erroneous judgment based on certain resemblances when we judge that animals have souls. But the stranger also makes a judgment based on resemblance—a resemblance *raisonnée,* however, based on his understanding of the workings of animal-like machines. The underlying reasoning in part repeats that of *L'Homme* and the *Description*: the properties I attribute to the automata whose fabrication I witnessed are, I now see, sufficient to explain all the operations of these so-called "animals," which the people around me, by a strange but inveterate custom, endow with souls they say are something like their own. To that reasoning are added the criteria of the *Discours,* which alone would compel the stranger to regard those animals as something other than very complicated machines. He carries with him the memory of the range of operations of which his automata were capable, and, seeing that the range is not *essentially* greater in animals, judges that, however they are composed, they

[9] The distinction is made in a letter to Mersenne on flying machines: "One can certainly build a machine that holds itself up in the air like a bird, metaphysically speaking [*metaphysicè loquendo*]; birds themselves, at least according to me, are such machines; but not physically or morally speaking [*physicè* ou *moraliter loquendo*], because it would take springs so subtle and overall so strong, that they could not be made by men" (To Mersenne 30 Aug 1640, AT 3:163). 'Moral' refers to practice, and here could well be translated as 'feasible.' On this passage, see Gontier 1991:5–6.

could be simulated by automata resembling his in all the respects that matter: there is nothing in them whose properties are not those he attributes to the parts of machines—a list that includes the modes of extension, but not faculties or forms. (After all, the argument loses its force if in his homeland the stranger had attributed a "horodictic" faculty to watches, or a "hydromotive" faculty to pumps.) The stranger, unlike us, is not deceived, because unlike us he has always regarded the nonhuman self-movers around him with the eye of an engineer.

The first twist in Descartes' treatment of the illusions produced by automata is that in fact we cannot be deceived by animal-machines, since animals *are* machines. At most a machine-bird of human manufacture would deceive us into thinking it was a machine-bird of divine manufacture, which is like mistaking a counterfeit Rolex for the real thing. The second twist is that we are, nevertheless, deceived by the animal-machines around us, because in the stage of life at which we first encounter them we know about ourselves but not about machines. For lack of better analogies, we take them to resemble us, not only in their operations, but in the principle of those operations. We think they have souls. The stranger, who has lived among machines (and only among machines) all his life, is not tempted by the *bad* resemblance of animals to us, but draws instead on the *good* resemblance between machines and animals. He is not tempted to attribute souls to the machines around him because he knows, having built some himself, *exactly* how they work.

5.3 The Scope of Intention

Through this tale two threads can be traced. The first, which I have been tracing until now, is that of resemblance: good resemblance, controlled by reason, which leads to truth; bad resemblance, now or formerly under the sway of the senses and the passions, which leads to error. The other is that of intentions: intentions to deceive, or to make something that resembles another. To them I now turn.

Is all the machinery of imitation necessary to a Cartesian science of life? An odd slip in the response to Pollot suggests that it is not. The stranger is said to have been raised "in a place where no animals but men had ever been seen" ("où il n'auroit iamais veu aucuns autres animaux que des hommes"). Yet of the automata he helps build, "some have the figure of a man, others the figure of a horse, others of a dog" and so forth. But how was it, if no animals had ever been seen, that some had the figure of a horse or dog? Was this a coincidence? God, in building the statues of the *Traité de l'Homme,* no doubt knew what we look like. But the stranger has only a human knowledge of the things around him. Was the resemblance then a mere accident?

What would the stranger come to believe if the automata built with his help had *not* had the figures of animals? In general, simulation does not require likeness of outward appearance. A simulation is adequate if the behavior, or as Descartes would say, the movements of the simulator resemble those of the thing simulated. A model of the arm might be made of sticks and springs, if all that is to be simulated is its lifting capacity. In Descartes' arguments for the animal-machine, however, it is difficult to imagine how something that *looked* very different from a dog could be said to imitate its actions perfectly: how would we set up a correspondence between the one set of actions and the other? You can't lick something without a tongue, or wag your tail if you don't have one. Since in several passages (e.g., the response to Pollot), Descartes holds that the expressions of animals can be imitated precisely too, one must suppose that the dog-machine looks very nearly like a dog. Otherwise it would be quite unclear what likeness of expression meant (it cannot, of course, mean 'likeness of sentiment expressed').

The stranger might not have arrived at any firm view about our animals if the automata of his homeland were clocks and pumps. Coincidence in figure, conundrum that it is, is necessary if the example is to be persuasive, since only by that coincidence can resemblance in movement be determinately judged.

One might ask: why resemblance only in figure and movement? Why not in feeling and thought? A letter to More asserts that

> it seems agreeable to reason that, since art is the imitator of nature, and men can fabricate various automata, in which there is motion without any thought, so nature also produces its automata, far exceeding those made by art, namely all the *bruta*.[10]

Reading this, one might reply: there is art and art, and alongside the art of figural imitation there is an art of dramatic imitation. Why should the same argument not work in the case of an actor who attempts to think like Reynard the Fox, or feel the timidity of the sheep? Why should figural art alone matter? Perhaps nature "impersonates" the divine, each creature in its own fashion, as the actor impersonates a character.

---

[10] It would be misleading not to quote the next clause, which strengthens the argument: "especially since we recognize no reason on account of which, given such a conformation of members as we see in animals, thought must also be in them" (To More, 21 Feb 1649, AT 5:277–278). The second clause provides the grounds for identifying *bruta* and the automata that Nature is argued to be capable of making.

A letter to Gibieuf suggests an answer. We observe in animals "movements similar to those that follow upon our imaginations or sensations, but not, for all that, imaginations or sensations" (To Gibieuf, 19 Jan 1642, AT 3:479). In the thought-experiment of the *Discours,* when we try to determine whether the creature before us is machine or man, what we observe is not the use of speech exactly, but the production of sounds (Descartes uses the expression 'proférer des paroles,' which focuses on the physical in speech). The actor, then, does not know what she is attempting to imitate: even if she is of the opinion that she imitates thoughts and feelings, she can with reason imitate only the cry an animal makes, the shape of its countenance.

But on what is that restriction on the observable based? It is true that in the physics of the inanimate world, there are no thoughts or feelings, but only modes of extension. Descartes argues in the *Principles,* even if there were such entities as colors and sounds, we could not perceive them except through the movements they might cause; but a body causes movement only by way of its modes of extension. Hence we could not perceive them at all.[11]

Grant that conclusion and go on.[12] It would seem that, as long as the physiology cannot do without models that exist only in the mind—the grand thought-experiments of *L'Homme* and the *Description*—then simulation is

---

[11] We "best understand how by the varying size, figure, and motion of the particular [parts] of one body, various local motions in another are excited; while in no other way can we understand how from them (size, figure, and motion) something else can be produced, entirely diverse in nature from them, such as those substantial forms and real qualities that many [philosophers] suppose are in things, nor yet how those qualities or forms might have the power to excite local motions in other bodies. That being so, and since we know that it is in the nature of our soul that different local motions suffice to excite all its sensations [*sensus*], and we experience them exciting various sensations in it, but do not grasp anything else, except motion of that sort, moving from the external senses to the brain: we must conclude altogether that those things which in external objects we pick out by the names 'light,' 'color,' 'odor,' 'taste,' 'sound,' 'heat,' 'cold,' and other tactile qualities, and also 'substantial form,' are not anything other than various dispositions of those objects which bring it about that they [sc. the objects] can move our nerves in various ways." (PP 4§198, AT 8/1:321–323).

[12] One might well wonder whether the argument does not prove too much. Other people can act on our senses only by way of their bodies; and by parity of reasoning with forms and qualities, we have no way of understanding how from acts of will something else, namely bodily movements, can be produced which is entirely diverse in nature from them. What tells against forms therefore tells against minds. Descartes' answer is that we know from our own acts that minds *can* move bodies, and the two criteria mentioned above for distinguishing automata with souls from automata without souls will enable us to infer that in some of the automata around us there are souls. But the Aristotelian might well reply: yes, but in my natural philosophy the human soul is a substantial form, and so I understand too how a substantial form can produce local motion. Some substantial forms, however, are different enough from human souls that not all the powers of the human soul need be attributed to them; contrary to what some philosophers seem to believe, the Aristotelian is not, merely by virtue of employing the vocabulary of forms, an "animist"; what belongs to all forms is not thought or will but power.

inescapable. The only criterion of success in a thought-experiment is that the phenomena imagined to result from the imagined machine resemble those we observe, which are modes of extension alone, or temporal sequences of modes—the cycle of movement and change of shape of the heart, for example. The imagined machine, therefore, is inflected from the start with the intention to resemble something we know by observation, a human body in action, say.

The scope of the required simulation is reduced somewhat, however, by the analysis of capacities, the breakdown of the machine into mechanisms, and the multitude of comparisons by which inferences from imagined mechanisms to their operations are justified. The comparison itself not only justifies the inference, it also suggests a description of the observed motions of the organ with which the mechanism is to be compared. For example, the body sweats and secretes various other fluids. We learn that in the man-machine

> there are also [particles of blood] that flow out into the urine through the flesh of the kidneys or into sweat and other excrements through the skin. And in all these places, it is only the situation or figure or smallness of the pores through which they pass that makes some pass rather than others, and prevents the rest of the blood from following them: as you may have seen different sieves which, being differently perforated, serve to separate different grains from each other. (*L'Homme*, AT 11:127–128)

Excretion is just one, not very important, action in the ensemble that together characterize the circulatory system. We are to imagine a series of filters which, like the much larger sieves of everyday life, differ by the size and shape of their holes, allowing some particles to pass through but not others. The phenomenon of sweating, on the other hand, is neutrally described as something like the appearance on the skin of a fluid resembling saltwater. We never see such a description in Descartes. Instead sweating is *already* a passing of something *through* the skin (unlike dew, a deposit of something *on* the skin, or melting, an alteration in the qualities of a heated solid). That enables it to be brought forward for comparison with sifting.

The global intention to simulate all the actions of an animal or human body is atomized into a series of more modest intentions to simulate first this kind of action, then that (the kinds being taken largely from pre-existing classifications of bodily phenomena). Correspondingly, the objects that simulate are mechanisms or parts of mechanisms, not the whole machine. The advantages of proceeding in this way are obvious. Convincing comparisons, drawn from inanimate nature, are much easier to find for restricted classes of behaviors.

The implicit promise is that the "atomic" capacities that correspond to those classes can be fitted together into "molecular" capacities, like the cycles described in §1.1, and finally into an explanation of everything the body does. The implicit danger is that the body will cease to be anything more than a collection of mechanisms that lack any but the *de facto* unity of being contiguous. We would be left in Baglivi's machine shop.

[6]

Unity of the Body

For common sense and the schools, an animal is a unity, a paradigmatic individual substance. It would seem that the Cartesian is bound to disagree. The machine has no special status among the configurations of matter. It is an enormous collection of parts of matter, joined locally by relations of contiguity, common motion, or momentary contact. But the arrangements we call animals and plants are no more natural than other arrangements or none at all. Descartes insists time and again that nature's laws are indifferent to the distinction between a living, intact animal and a mutilated, diseased, or dead animal.

Yet he, like everyone else, treats animals and plants and their organs as phenomenal unities. Does he stick to the letter of his doctrine, and treat those unities entirely on a par with clocks? Can he do so consistently while putting forward his version of the science of living things? The answers, we will see, are not straightforward. In what follows I first outline what might be called Descartes' official view on reasoning from ends. I then develop four conceptions of unity applicable to collections of simple bodies (their individuation, troublesome though it is, I here take as given). *Physical* unity is the unity that parts of matter have when they share a common motion; *dispositional* the unity of something whose parts are said to be "disposed" or arranged in such a way that the whole has some designated effect; *functional* or *intentional* the unity of a thing which has the purpose of producing a designated operation (in Descartes' way of thinking, 'functional' and 'intentional' coincide, as we will see); *substantial* the unity which a collection of parts of matter enjoys by virtue of being joined with a single substantial form. In Descartes' world only the human body, when joined with the soul, has substantial unity. For that reason substantial unity lies largely outside the scope of this work, and my treatment will be brief.

6.1 Against Ends

In the vocabulary of Aristotelianism, *functio* is allied with terms like *officina* and *munus,* which in their most direct usage denote roles or duties.[1] On the face of it, Descartes has no use for such notions in his natural philosophy. The stark rejection of any appeal to ends is exemplified in a paragraph from the *Principles:*

> we do not avail ourselves of any reasonings [*rationes*] concerning natural things [taken] from the end which God or nature in making them proposed to itself [added in French: and we reject entirely from our philosophy the search for final causes]. We are not so arrogant as to think ourselves party to his deliberations. Rather, considering God as the efficient cause of all things, we will see what, from certain of his attributes which he has wanted us to have some knowledge of, the natural light which he has imparted to us discloses that we ought to conclude concerning certain of his effects that appear to our senses.[2]

Descartes' constant refrain is that, while reasonings about particular divine ends may have a role in ethics or apologetics, they are "inept" in physics. We would, moreover, be overstepping our limits if we inquired after them, the implication being that to inquire into divine purposes puts one only a step away from questioning them.[3]

---

[1] 'Office' occurs in a letter to Élisabeth: "l'office du foye & de la rate est de contenir tousiours du sang de reserue" ("the office of the liver and spleen is always to contain blood in reserve") (To Élisabeth, May 1646, AT 4:407).

[2] *PP* 1§28, AT 8/1:15–16; 9/2:37. I have consulted CSMK for help in translating the second sentence. The French transfers the phrase 'from certain of his attributes,' etc. to another sentence about the guarantee given to clear and distinct ideas which is absent from the Latin. A letter to Élisabeth makes the point very clearly: if we "attribute to them other imperfections they don't have, in order to raise ourselves above them," we engage "an impertinent presumption," and "take charge of the world with [God]," which "causes an infinite of useless inquietudes and vexations" (To Élisabeth, 15 Sep 1645, 4:292).

[3] To ****, Aug 1641, AT 3:431, and the letter to Élisabeth just cited. It is worth noting that Descartes always refuses to admit in science theological grounds for asserting human preeminence or dominance over nature (see, for example, his interpretation of the days of creation: To Chanut 6 Jun 1647, 5:53–55). Revelation aside, whatever mastery we assert, we assert for ourselves, without the sanction or command of God. In particular, no policy regarding the treatment of animals can be inferred from the fact that they are machines, though we will indeed have no duties toward them as we do toward humans, if those rest on having souls.

His practice, by and large, conforms to his precept. The physiological texts are striking from that standpoint. In other texts of the period, owing to the precedent not only of Aristotle but of Galen's masterwork *De usu partium,* ends and final causes are ubiquitous.[4] Even Gassendi, no friend of Aristotle, argues at length in rhetoric reminiscent of Galen's that the parts of animals testify to the sagacity and industry of the creator.[5] In a paragraph whose targets might well include his illustrious contemporary, Gassendi writes:

> To us it suffices that no one can, with open eyes and attentive mind, fail to acknowledge immediately that any person must be bereft of reason who believes that the parts of Animals were elaborated without reason, deliberation, foresight, and appointment to certain uses and ends. (Gassendi *Syntagma* 2§3mem2lib2c3, *Opera* 2:231a)

Adducing *all* the usual instances on behalf of the role of divine foresight in the constitution of plants and animals, he concludes that

> just as the Divine Performer of works [*Divinus tamen Opifex*] did not require pre-existing matter to operate, so too in directing things he required no external exemplar; for by his immense insight he understands no less those things which are not than those which are, and from the fecundity of his own intellect, both conceived the work that was to be, and foresaw its aim. (Gassendi *Syntagma* 2§3mem2lib2c3, *Opera* 2:236b)[6]

It is not surprising that the strictures in the fourth *Meditation* against appeals to ends in natural philosophy should have caught his eye. Descartes there argues that since God is immense, incomprehensible, infinite, "that whole genus of cause, which customarily argues from ends, I judge to have no use in Physical matters," adding the usual warning about temerity (*Med.* 4, AT 7:55).

Those strictures no doubt apply, Gassendi agrees, to ends "which God himself has wished to be hidden, or of which he has prohibited the inves-

---

[4] See Bitbol-Hespériès 1998 for a survey of medical texts in which God's ingenuity and goodness is praised.

[5] On the integral role of divine ends in Gassendi's natural philosophy, Bloch writes, paraphrasing Gassendi: "If [. . .] physics is a contemplative or speculative science, it is that the world and all it contains are a divine work, the product of a divine art [. . .]; all our science can do [. . .] is contemplate in natural phenomena the works of an art and wisdom which are not ours" (Bloch 1971:435, see also 71, 359, 434–439, 444).

[6] The reference to "exemplars" is, I take it, directed against Platonists, not against natural theology.

tigation." But not to those "which he has set, as it were, in an open space[7] and which without great labor become known" (4 *Obj.*, AT 7:309; Gassendi *Opera* 3:359a). To which Descartes replies, with all the air of one who has arrogated the title of 'philosopher' to himself (4 *Resp.*, 348–349, 352), "One cannot pretend that some of God's ends, rather than others, are out in the open; all are hidden in the same manner in the inscrutable abyss of his wisdom" (375).[8]

All of God's ends hidden in the abyss! Had Gassendi not studiously followed the order of topics in the *Meditations,* he might have mentioned the passage in the sixth *Meditation* in which Descartes affirms that the relations instituted by God between the motions in the brain and the sensations they give rise to are, "of all that could be inferred, those which are most, and most often, conducive to the conservation of human health" (*Med.* 6, AT 7:87). What is human health in this passage but a divine end? How indeed might one appeal to God's goodness if not by considering his ends?[9]

It is difficult to see how Descartes could avoid appealing to ends in the vindication of the senses in the sixth *Meditation.*

First, with respect to the understanding. Divine veracity is inferred both from God's goodness and from his power. God is not malicious, and so he would not deliberately make us believe that the cause of our sensations is extended substance if it were not. God's intellect and will, moreover, are not so weak (*imbecillus*) that he would give us the power of being able to form judgments capable of truth and then neglect to ensure that the conditions under which the power may successfully be exercised actually obtain.

Second, with respect to the senses themselves. If they are to have any use in the search for truth, the sensations that result from them must be such that a true theory, not merely of the most general features of the natural world (namely, that it consists of *res extensa*), but of particulars (like the size of the sun) should be attainable by their use.[10] It might well be consistent with their

---

[7] '*In propatulo*': *propatulum* denotes in particular the forecourt of a house, the portion that can be seen from the street. We might say: in the shop-windows of the world.

[8] A letter of August 1641, written in response to another set of objections to the *Meditations,* even affirms that it is "per se manifest" that the ends of God cannot be known (AT 3:431).

[9] Where the *Meditations* alone are concerned, the answer turns on two points: (i) whether every appeal to God's goodness can be reduced to an appeal to his veracity; (ii) whether the argument on behalf of divine veracity covertly enlists ends. "In every falsehood or deception some imperfection is found"; and God, evidently, cannot fail to know the truth; on the other hand, he is neither stupid nor malicious. The crux is malice. The authors of the second *Objections* note instances of what would seem to be benevolent deception in the Bible. Descartes effectively denies that these are instances of deception, thereby avoiding the consideration of the "internal and formal malice" which deception properly consists in (AT 7:143).

[10] "As for the rest [of the things that might be known], which are either only particular— that the sun has such-and-such a size and figure, etc.—or else less clearly understood, like light,

role in the conservation of health that the senses should fail to represent bodies as they really are, it is not consistent with the possibility of knowing particulars that they should yield no information at all about the external world.

Even where he is discussing the body alone, Descartes often gives the appearance of appealing to purposes. *L'Homme* is bracketed between two occurrences of the word *fonction:*

> I suppose that the Body is nothing other than a statue or machine of earth, which
> God forms expressly to make it as similar to us as possible [. . .] and finally that
> it imitates all the functions that can be imagined to proceed from matter, and to
> depend only on the disposition of the organs. (*L'Homme,* AT 11:120)

∽

> I wish you to consider [. . .] that all the functions I have attributed to this Ma-
> chine, like the digestion of food, the beat of the heart and the arteries, the nour-
> ishment and increase of its members, breathing, waking and sleeping, [a long list
> follows] [. . .] I wish that you should consider that these functions all follow
> naturally, in this Machine, solely from the disposition of its organs. (*L'Homme,*
> AT 11:202)

One might suppose that in these passages 'function' is a mere *façon de parler,*
and that 'movement' would do as well.[11] But Descartes on many occasions
uses what grammarians would call 'clauses of purpose.'

----

sound, pain, and so forth, although they are very doubtful and uncertain, from this fact alone,
that God is not deceptive, and so it cannot be that any falsity should be found in my opinions,
unless there is some faculty bestowed on me by God to correct it, I gather some hope of grasp-
ing the truth in them also" (AT 7:80). The argument rests on no other basis than that the
"complexion" of my cognitive faculties has been intended by God to be capable of truth—or
at least of avoiding error. But it seems clear that in the instances here mentioned the "correction" (*emendatio*) of error does not consist merely in the suspension of judgment. There is a very
careful version of the argument in a late letter to Clerselier. A being in which there is no imperfection cannot tend toward non-being (*non ens*), or "have non-being for its end and institution"; but falsehood is non-truth (*non verum*), and so a being without imperfection cannot tend
toward falsehood (To Clerselier 23 Apr 1649, AT 5:357). Any situation in which a creature of
God is, by its own nature or the nature of things, bound to fail, is a situation in which God himself, in the act of creation, must have had non-truth for his end; but since God is perfect, that
cannot be.

[11] Cf. *Discours* 5, AT 6:46.12. A parallel passage in the *Description* does have 'movement'
rather than 'function': "tous les mouvements que nous n'experimentons point dépendre de nos-
tre pensée, ne doivent pas estre attribuez à l'ame, mais à la seule disposition des organes" (AT
11:225).

And the fire in the heart of the machine [. . .] does not serve for anything else but to dilate, heat, and subtilize the blood. (*L'Homme,* AT 11:123)

∞

you can easily also understand how this machine may sneeze, yawn, cough, and make all the other movements necessary to reject various other excrements. (141)

∞

But the little threads that compose the marrow of the nerves of the tongue, and which serve as the organ of *taste* in this machine, can be moved by actions slighter than those nerves that serve only for touch in general. (145)

The spleen is "destined to purge the blood" of certain particles, the kidney "destined" to purge it of others (169); if the hand is heated too much by fire, some of the animal spirits are conducted to the nerves "that serve to move the exterior members, in the fashion required to avoid the force" of the fire's action (193). The passions "serve to dispose the heart and liver" and so forth "in such a way that the spirits that are generated then turn out to be proper for causing the movements that ought to ensue" (i.e., movements to avoid harm or pursue benefit) (193). In the *Description,* Descartes writes of the "principal use [*usage*] of the lung" (AT 11:236). All the branches of the great artery are joined to branches of the *vena cava* "in such a way that after having distributed to all the parts of the body the blood they ought to receive, either for the nourishment, or for other uses," they bring the rest to the *vena cava* and the heart (238). The dilatation of the blood in the heart changes the nature of the blood "as much as one might imagine it ought to be changed, in order that the blood should be prepared, and rendered more proper to serve as nourishment for all the members, and to be employed in all the other uses it serves in the body" (244). This in a passage where he is arguing against people who attribute "faculties" to the heart in order to account for changes in the quality of the blood. Although such constructions occur far less frequently than in Aristotelian works (especially in the description of the formation of the fetus, where they are sparse indeed), and although some (like 'so disposed that') are ambiguous, there are sufficiently many that one cannot regard them as mere slips or as concessions to the vulgar way of speaking.

If we follow the official view, it would seem that only two principles of unity will be available: physical and dispositional. Functional and intentional principles have no role. They require a knowledge of divine intentions that in natural philosophy we cannot attain to. But it remains to be seen whether the official view is a sufficient characterization of Descartes' stance.

6.2 Physical Unity

A principle, according to Aristotle, is "the first thing, from which something is, or comes to be, or is known" (*Meta.* 5c1, 1012b33–35). Suárez reduces the list to two, the first of which, the *principium ejus quod fit,* the principle of that which comes to be, includes "every principle of change or operation, as such, or even of any successive thing." More pertinent here is the *principium ejus quod est,* the principle of that which is (or exists), which includes, for example, the material and formal causes of a complete substance: these do not bring about change, each is in its own manner that by which a thing is (or is such-and-such, in the case of form). 'Principle' is broader than 'cause,' because entities of reason and privations have principles but not causes.[12] A *principle of unity,* then, is the primary ground of a thing's being one. A living body, evidently, has parts—its organs, the blood—which, even if they are not distinct substances, can subsist in some way (though perhaps not as organs or blood) apart from the rest. The principle of its unity is the ground upon which the body is nevertheless rightly said to be one thing.[13]

The most basic principle of unity in Cartesian natural philosophy is mutual rest (*PP* 2§55, AT 8/1:71). What unites the left half of a second-element sphere with the right half is simply that they are not moving with respect to one another. According to Descartes' rather peculiar definition of movement,

---

[12] Suárez *Disp.* 12§1no11, 12, 25, *Opera* 25:377, 382. There is a brief discussion of principles in a letter of Descartes to Clerselier, but the principles in question are principles of knowledge, not of being or becoming (To Clerselier, Jun or Jul 1646, AT 4:444).

[13] '*Individuum,*' as its etymology indicates, denotes something whose parts are either incapable of separate existence as something of that kind, or which has no parts. But the problem of individuation is typically presented, perhaps in reminiscence of Neoplatonic doctrines of emanation, as one of "dividing" a species into its members. If Peter and Paula had no mode or accident but those they have by virtue of being human, they would be *one*. A "principle of individuation" is whatever it is that, added to the specific nature which is exactly the same in both, makes them *two*. If we consider a human being in its existence as a complex of variously distinguishable parts, then its principle of unity is whatever justifies saying of each part that it belongs to *this* whole (and of alien parts that they do not). Since parts, to be identified, must be individuated, individuation is prior to unity. On the other hand, a principle of unity does permit one to distinguish individuals by distinguishing the parts of one from those of another. For Suárez, individuality is equivalent to the "individual and singular unity" possessed by each member of a species (*Disp.* 5§1no7, *Opera* 25:148). Form is in most cases an adequate principle of individuation for complete substances; being joined with form yields a principle of unity for the proximate matter of a substance. (On all these matters, see Gracia 1994a.)

that means that at this moment they are not rupturing, or not ceasing to share a common boundary, like ice sheets slipping off the face of a glacier. That principle has its difficulties, but they do not impinge greatly on physiology.[14]

The other principle of unity inherited by the physiology from the physics applies primarily to collections of individual bodies of the fundamental sort. In the definition of 'one body' (*unum corpus sive una pars materiæ*), Descartes specifies that a body is "everything that is transferred at once, even if this is constituted from many particles that in themselves have other motions" (*PP* 2§25, AT 8/1:54). A ticking clock carried to the prow of a ship, or the earth with its oceans and its proper vortex, can be said to be one body because the whole collection moves at once with respect to the ship or to the sun and the fixed stars (2§31, 8/1:57; 3§29, 8/1:92).

Again this less strict notion is not without its difficulties (Grosholz 1991: 68–69). Set aside the difficulties that pertain to animate and inanimate alike. Where plants and animals are concerned, the definition draws some odd boundaries. The air in my lungs, the water I have just taken into my mouth, will be part of my body. Two lovers locked in tight embrace will, it seems, really be one flesh. It does seem to be true, on the other hand, that the body of an animal will be *included* in a single body: the problem is that too much else will be also.[15]

But the chief objection to physical unity applied to organisms is that it has nothing to do with their nature as organisms, as *living*. It would be strange if the nature of an organism, and the conditions of its unity, were so divorced.[16] Descartes recognizes that point where the human body is concerned.

> Although the entire mind seems to be united to the entire body, nevertheless when a foot or arm or whatever other part of the body you please is cut off, I recognize nothing to have been thereby removed from the mind. (*Med.* 6, AT 7:86)

---

[14] See Grosholz 1991:63–71, Garber 1993:175–181, Des Chene 1996:374–382, Wilson 1997.

[15] The definition is too narrow to accommodate Christ's scattered body in the attempt at explaining the real presence that Descartes sent (with some misgivings) to Mesland in 1645 or 1646 (AT 4:345–348). Since the explanation depends on the notion of substantial union, I will discuss it under that heading.

[16] Allowing, for the nonce, a family to be an individual of a sort, the point is that it is one *family*, not one collection of *bodies*, and that its unity as a family ought to be derived from its nature as a family, not its nature as a collection (or mereological sum) of bodies. The contemporary notion of supervenience would seem to allow for violations of the maxim. If the mind supervenes on the brain or the body, then questions about the individuation of minds devolve onto questions about the individuation of brains, since there can be no difference in mental properties except if there be a difference in physical properties. Mental properties per se are irrelevant to determining the individuals in this world.

It is no less a mind after than before: which is to say, its unity does not depend on the integrity of the body, and that is because its essence as a thinking thing has nothing to do with body. But a dog, too, though its tail be bobbed, remains no less a dog than it was. Or again, if we imagine that part of the human body—by the resection of nerves, say—no longer is united with the soul, no more so than a prosthesis, then the thing which is physically one and the thing which is joined with the mind are no longer the same. So too a tree one of whose limbs has died is physically one thing with the limb, but as a living thing it is only what still bears leaves, to which some wood happens to be joined.

Everything in the Cartesian world of nonliving things is a "part of matter," a region of extension delineated by the occurrence of rupture at its boundaries. But among the parts of matter there is a great variety of shapes and sizes, which correspond to commonsense kinds like water, oil, glass, iron. Descartes has no uniform term for what distinguishes those kinds. Malebranche, noticing the absence, uses 'configuration.'[17] The configuration of blood, for example, consists in its having numerous twisted branches, which is "proper" to blood. All and only particles of blood have that configuration. *In each instant* its unity as a particle of blood and its physical unity amount to the same thing: it will be one blood-particle just in case it is one body. But if we consider its unity from one instant to another, it can cease to be one body without thereby ceasing to be one blood-particle. Descartes supposes that as they circulate through the body, blood-particles tend to be pruned of their branches. Yet the same particles pass through the heart many times.

The strict criterion of physical unity, then, does not coincide even with criteria of unity for inanimate kinds like oil or blood (considered in its own right, as a viscous sticky fluid); still less is it sufficient for the unity of machines. Neither is the lax criterion. In analyzing the capacity of a pump, Caus isolates mechanisms—the system of gears and cylinders, as opposed to the gear train that transmits the motion of the horses to that system. The parts of a mechanism, as it operates, move in many directions. That its parts should, in what we would now call a certain inertial frame, all have the same uniform motion is relevant to its unity as a mechanism, but only in the following respect: if the parts did not satisfy that condition, then in time, some parts whose contiguity is necessary to the functioning of the machine would cease to be contiguous, and the machine would no longer have the same disposition or

---

[17] "Extension is capable of receiving two sorts of figures. Some are merely exterior, like the roundness of a piece of wax; others are interior, and it is these that are proper to all the little parts of which the wax is composed [. . . .] I call simply *figure* that which is exterior, and I call *configuration* the interior figure, which is necessary to all the parts of which the wax is composed in order for it to be what it is" (Malebranche *RV* 1CI§1, *Œuvres* 1:23).

fulfill its function. In short, physical unity matters only because it is necessary to disposition or function. Dispositional or functional unity is in that sense prior to physical unity. The Aristotelian accepts that conclusion wholeheartedly. Dispositional unity is exactly that unity which is conferred on the combination of matter and form by its form, from which its powers flow. The question, then, is whether Descartes is compelled to admit dispositional unity into his natural philosophy, and whether that is consistent with his principles.

6.3 Dispositional Unity

The term *dispositio,* as I have mentioned, means 'an arrangement to some purpose or end.' How much weight is put on *arrangement,* and how much on *purpose,* varies with context. Arrangement, in the cases of interest here, denotes spatial arrangement. Spatial arrangements are disrupted if their parts do not maintain, in some broad sense, more topological than geometric, their mutual relations in space. To that extent the lax criterion of physical unity would seem to suffice. But the more weight put on purpose, the less relevant that criterion becomes.

Descartes' use of *dispositio* is sometimes tantamount to 'whatever arrangement can operate in the specified way.' The process of digestion consists in part in the filtering of food particles through sieves of various sorts. Of them Descartes writes: "As for the size and figure of the pores [in the filters], it is evident that it suffices to bring it about that particles of blood which have a certain size and figure will enter into certain locations in the body rather than others [. . . .] Thus the blood, pushed by the heart into the arteries, no doubt finds various pores, through which some of its particles can pass and not others" (*Descrip.* 3, AT 11:251). It hardly matters what the size and figure of the pores actually are, but that some such arrangement should be capable of the task of separation, so that faculties of selection and attraction are superfluous. Similarly, in the formation of the organs of sense, or the conditions of the animal spirits that give rise to the passions, the details are, in a way, gratuitous, so long as it is plausible some arrangement or other can produce the required operations. In this case we can say a little bit more—namely, that different effects must have different causes—but not much. The rest is invention.

Dispositions and functions alike share a more or less marked indifference to matter, to the particulars of realization. But in early modern philosophy disposition maintains its long-standing ties with arrangement and order. In this century those ties have been loosened. Instead it has taken over some of the territory of the older, disused terms 'faculty' and 'power'; its now dominant use outside philosophy is to denote temperaments or habits of mood. The notion of arrangement, if present at all, is off in the wings.[18]

Descartes' laws of motion are stated in the indicative; the rules of collision as counterfactual conditionals.[19] Each of the rules describes what we might think of as a simple arrangement of parts of matter, and then what motions would (or always do) result from it. Describing the motions of a clock is only a more complicated, not a different, task. To say that the motions of the body, or its functions, "follow from" its dispositions is to sum up an indefinitely large collection of propositions of the form 'if the body were acted upon thus, such-and-such motions would occur.'

Function, on the other hand, focuses rather on the operations.[20] Descartes tends to use it in more programmatic contexts.[21] He uses it, moreover, of the operations of both body and soul.[22] *Dispositio* too can be used of capacities of the soul.[23] But Descartes seems not to use it that way, no doubt because the soul is not supposed to have parts that could be arranged. That difference in use survives despite the shift in use of 'disposition.' Sugar has the disposition, not the function, of solubility; there are functionalist, but (to my knowledge) no dispositionalist, philosophies of mind.

[18] It is not altogether absent: see Quine 1960:223.

[19] *PP* 2§46, AT 8/1:68. The French preserves the counterfactual construction (AT 9/2:89). Some English translations do not. I should note that Beeckman's early statement of some rules of collision, written in collaboration with Descartes, uses the indicative (Beeckman *Journal* 23 Nov-26 Dec 1618; 1:265–266). So perhaps no great weight should be put on grammatical mood. What matters is the character of the laws that the subjunctive imperfect conveys, namely, that they are "unreal" conditionals, which hold in conditions that need not obtain. Some philosophers have sought to reduce dispositions in the sense of '(active or passive) power' to counterfactuals: 'x is soluble' is supposed to mean, or be equivalent to, 'x would dissolve if x were immersed in the appropriate solvent.' The counterfactuals are in turn supposed to be backed up by general laws. In effect, the proposal defers the problem of distinguishing genuine dispositions from merely accidental properties (planets have no "disposition" to revolve around the Sun even if all known planets do) to that of distinguishing genuine laws from accidental generalizations. Descartes' treatment of relatively simple, but nonfundamental properties like flexibility and weight tends in the same direction; his laws, for their part, are genuine because derived from the immutability of the divine will, which in matter manifests itself as the preservation of the total quantity of motion.

[20] The verb *fungor* means to perform, execute, serve as; in most occurrences, what is performed is a duty. *Fungi officio* was a circumlocution for 'die.'

[21] As in the titles of three articles in the *Passions:* "Breve explication des parties du corps, & de quelques unes de ses fonctions," "Quel est le principe de toutes ces fonctions," "Quelles sont les fonctions de l'âme" (*PA* §7, 8, 13, AT 11:331).

[22] The soul is known to us only by the fact that it thinks, "that is, that it understands, wills, imagines, remembers, and feels, because all these functions are species of thought [toutes ces fonctions sont des especes de pensée]" (*Descrip.*, AT 11:224).

[23] *Dispositio* can denote the arrangement of virtues, for example, that together make up some other virtue. Prudence is deliberation, judgment, and command suitably ordered (Suárez *Disp.* 42§3no10, 12, 25, *Opera* 26:613).

The question now is whether dispositions provide a principle of unity, independently of their purposes. With mechanisms there are two possibilities.

1. Linkage

The basic unit in a machine is a linkage between its parts. Reuleaux, after a careful survey of previous attempts to delineate the class of machines, defines a machine as "*a combination of resistant bodies so arranged that by their means the mechanical forces of nature can be compelled to do work accompanied by certain determinate motions*" (Reuleaux 1874−5/1876:35).[24] One resistant body can determine the motion of another only through contact.

> In the machine [. . .] the moving bodies are prevented, by bodies *in contact* with them, from making any other than the required motions. This contact also, if the problem is to be entirely solved, must take place continually. (41)

Hence the machine "cannot so well be said to consist of elements as of *pairs of elements*" (43). Each element of a pair (except the first or last pair) must be rigidly connected with another. A linkage or "kinematic chain" is a sequence of such pairs, subject to the further condition that "*every alteration in the position of a link relatively to the one next to it be accompanied by an alteration in the position of every other link relatively to the first*" (46). Only then will every link have a determinate relative motion relative to the others. Reuleaux calls such a chain a "constrained closed" or "closed" chain. The simplest closed chain has four pairs, each consisting of a pin and collar—in Figure 14, *ab* or *cd,* for example. A link in the chain joins two pairs—*bc* or *de.* The possible paths of *all* the elements in the chain are determined when one link is fixed—by fastening the rod between *a* and *h* to a support or stand, say.[25]

The machine, then, requires continuous contact between pairs of parts, and its configuration is such that the movement of one part determines all the rest. That determination is geometrical, not mechanical: the laws of motion have no role in calculating the path of any part. In that sense, the operations of a machine, like the curves traced by Descartes' proportional compasses, follow precisely from its disposition.

We have already two kinds of unity in a linkage. The first is the connectedness of its parts. Between any two points in a linkage, no matter what its

---

[24] I have substituted italics for Reuleaux's (and his translator's) *gesperrte* (spread) lettering.

[25] The support or stand itself, though sometimes called a passive element, is not part of the machine at all (Reuleaux 1874−5/1876:48). Animals (except those that, like barnacles, fasten themselves permanently to something) have only temporary supports—the ground, a tree limb, another animal. The mechanisms within them, on the other hand, typically have permanent supports (the lower jaw, for example, acting as a lever, has part of the upper jaw as its fulcrum).

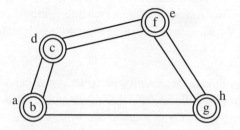

Figure 14: Kinematic chain (adapted from Reuleaux 1874–5/1876, Fig. 10, p47).

present state is, a continuous curve may be drawn. When it is stationary, it is in fact *physically* unified, if we neglect the possibility that the two elements of a pair may be separated by first- or second-element particles. Even when it is moving, the two elements of a pair cannot separate entirely: their reciprocal movement is a kind of sliding.[26] The second kind of unity is the mutual determination of their motions. In Figure 14, once one link is fixed, the motion of any other link determines that of all the rest.

It might seem that the hydraulic mechanisms that predominate in Descartes' physiology cannot be subsumed under Reuleaux's definition. But Descartes tends to think of unidirectional flows as if their kinematic properties were essentially those of rigid rods. In the *Dioptrique,* the traditional analogy between seeing something and touching it with a baton is applied to the more or less fluid medium of light.[27] The hydrodynamics of the pineal system are best understood if one treats the animal spirits issuing from the gland on the analogy of the constant outpouring of particles from the Sun. A "ray" from the gland is then not unlike a rigid rod wedged between the gland and the walls of its chamber in the brain, so that when a pore in one of the walls opens, there is a release of pressure on a particular point of the surface of the gland.[28]

The gland is "held up, as if it were in a balance, by the force of the blood that the heat of the heart sends toward it" (AT 11:179). One of Caus's *jeux d'esprit* offers a compelling analogy.[29] The second problem of the second book of his work gives the "design of a grotto where there is a Ball which is raised by the force of water."

[26] As in the motions of the "striated" or "channeled" particles that cause magnetic phenomena (PP 3§87, 4§139 & 146, AT 8/1:143, 279, 287). See also Des Chene 1996:265.

[27] Diop. 1, AT 6:86; 7, AT 6:135; cf. *Le Monde* 14, AT 11:100 and *L'Homme,* AT 11:160 & Fig. 15.

[28] *L'Homme,* AT 11:174–176. The *Passions* are much vaguer (PA §17, 31. AT 11:341, 352); some commentators hold that Descartes changed his view, and that the animal spirits are no longer supposed to issue forth from the gland, or that Descartes retreated to a sage agnosticism on the matter (Kambouchner 1995:133–138; his thorough treatment includes a survey of literature on the question).

[29] As far as I know, the analogy has not been noted in the literature. Baltrusaitis notes the close resemblance between the fountain-figures described in *L'Homme* and certain of Caus's designs (Baltrusaitis 1984:59–66; see also Gaukroger 1995:63).

Figure 15: A ball dancing on a spout of water (Caus *Forces mouvantes*, Liv. 2, Prob. 2).

in order to make the water raise the ball well if the water proceeds out of a tank, it is necessary that the bottom of the tank should be at least twelve feet higher than the surface of the earth, and at most twenty-four feet, and the pipe from which the water issues will be as thick as the little finger, narrowing a little at the tip, and the end through which the water issues will be at the bottom of a vessel in the form of a funnel to receive the ball more easily when it falls, and to remove the water that falls into the vessel, there will be holes at the very bottom of it. The rocks [i.e., the walls of the grotto] will be decorated with animals made of natural shells fitted and glued together, which will emit water through little pipes that they have in their mouths, so that those jets may sometimes strike against the ball to make it fall; immediately it will rise again by means of the water that pushes it upward, and thus hopping back and forth it will be pleasant to watch. (Caus *Forces mouvantes* Liv. 2, Prob. 2)

One would have only to arrange somehow that water should issue *from* the ball, and *enter* the mouths of the animals on the walls of the grotto, to have the pineal system. In the plate illustrating Caus's machine, moreover, the jets ignore the force of gravity, and look for all the world like rigid rods. Only the splashing water at their ends betrays their fluidity.[30]

Hydraulic devices, in short, are much like devices whose parts are rigid, except that their "rods" are easily changed in size and direction, or added and removed (by opening and closing valves), and are able to transmit motion along curved channels. It is reasonable, though at some cost to the precision of Reuleaux's definitions, to transfer the terminology of linkages and mechanisms to them also. The continuity of the jet replaces that of the rod, and the basic pairs are replaced by jets and chambers, in which the pressure of a jet against the chamber causes it to move or change shape. It is, in fact, the changes of shape caused by the motions of the animal spirits that give the machine the capacity to acquire new dispositions, and thus new patterns of behavior—a capacity more difficult to realize in rigid mechanisms.[31] There is,

---

[30] Though Descartes' correspondences between the felt qualities of the passions and the mechanical properties of the animal spirits may well strike us as naïve or "primitive" (Gaukroger 1995:273)—and no doubt are, even when compared to Descartes' own explanations of other phenomena like the perception of distance—, still the technology upon which he bases his comparisons was not at all primitive. Caus's work and others like it represent the hi-tech of the early modern.

[31] Barrel-organs and the like, in which pitch-pipes were triggered by cylinders with pegs on their sides (Caus *Forces mouvantes,* Liv. 1, Prob. 28–33), had what we would call "read-only memories." Pascal's calculator and other similar devices could be thought of as including temporary storage registers, but I know of no attempt in the seventeenth-century to apply either that or barrel-organ technology to physiology.

nevertheless, nothing aleatory or chaotic about hydraulic devices: their operations too follow from their dispositions.

A mechanism has a unity more complex than physical unity. It may, perhaps, be physically united according to the lax criterion, but that, as Reuleaux's argument indicates, is a consequence of its being a mechanism (and thus a necessary condition). Instead the unity of the mechanism consists in the joint influence of its parts on one another, an influence analyzable into the transmission of force by contact from each link to the next.[32] Were there action at a distance (or if electromagnetic forces were brought into play), contact would no longer necessary. One could well imagine a version of Caus's dancing ball made with magnets—a more elaborate device, but working on the same principle, as Descartes' dancing statue.

2. Concert

The machine, in its simpler forms, is intended to produce a *single* motion. Its elements work in concert to produce that motion, whether it be lifting water, transferring a load, or producing a jet. "Every motion that differs from the one intended will be a *disturbing* motion, and we therefore give beforehand to the parts which bear the latent forces [. . .] such arrangement, form, and rigidity that they permit each moving part to have one motion *only,* the required one" (Reuleaux 1874−5/1876: 35). The reference to intention points toward intentional unity: it is the intention of the machine-builder that supplies the norm against which some motions are disturbances, and some not. But there is, one might say, a fact concerning the motion of the machine that has nothing to do with intentions: the clock's hands would turn, the animal-machine's heart beat, even if they had popped out of the void. It is, as Descartes insists, a matter of indifference to nature whether the clock is capable of turning its hands, or the heart of beating; but that does not gainsay the fact that when it *is* whole, it does what it's supposed to.

The fact that the parts of a machine act in concert to produce one motion cannot without further ado be enlisted to the service of dispositional unity. The *one* motion stands out only, it would seem, in relation to the intentions of its builder (or someone else's guesses about them). The turning of a clock's hands is itself the cause of further effects—movements of the air, a changing pattern of shadows on the clock face. The beating of the heart has effects

---

[32] The image of a *chain,* which is indeed the term used by Reuleaux, is compelling, but it should not be forgotten that in a mechanism with flexible or compressible elements a part might be acted on through several linkages at once.

throughout the body. From nature's standpoint, the choice of a terminus for the concerted motions of the parts is arbitrary. Similarly the choice of the initial, activating motion. To isolate the machine from its environs we must, it seems, either look downward toward physical unity or upward toward functional and intentional unity.[33]

Descartes exhibits no interest whatever in such problems. He never asks himself why the sun whose light acts on the eye is not part of the visual system. To be sure, it shines on other things too. But the fire in my cat's heart warms me just as the fire in my own heart warms me. Even so, some of Descartes' pronouncements about the animal-machine make the most sense if what he has in mind as a principle of unity is dispositional unity. If his *diktat* against ends were strictly enforced, then the animal-machine can have no more than dispositional unity, because, as I will argue shortly, functional and intentional unity are virtually inseparable in his thought. Nevertheless, if what may be called the boundary problem could be solved without appealing to intentions, dispositional unity would supply Descartes with everything he wants—a way of picking out the animate individuals of the natural world while referring to nothing beyond it.

6.4 Functional and Intentional Unity

The occurrences of 'function' in Descartes' physiology might well be done without. Descartes himself, as I have mentioned, in a passage in the *Description* paralleling the introduction of the animal-machine in *L'Homme,* substitutes 'mouvement' for 'fonction.' Yet there are numerous passages in which a mechanism is said to "serve" a power, especially a power of the soul, like vision, or to be "for" some operation. Those passages, as I have said, are not easily dismissed or translated into counterfactuals. In Aristotelianism, such talk is taken to designate ends and, under certain conditions, final causes.[34] Any nat-

---

[33] The difficulty of picking out naturalistically the appropriate point in a causal chain is a well-known difficulty for causal theories of content. See, for example, Dretske 1981:156–157.

[34] A *final cause* is not merely an end. It is an end which can be brought under the general notion of cause as that which "inflows being," where 'being' means either existence *simpliciter*, as in creation or generation (in generation, the matter exists beforehand, but not a complete substance), or existence in some manner, as in alteration where a cold thing, say, is made warm. The final cause fits under this heading insofar as the form achieved when a natural change reaches its *terminus* is intended by an agent—in the actions of nonrational agents, God—under the "formal reason" of the good. My thought, considered as a mode of thinking substance, or "materially" in the terminology of the *Meditations,* is the efficient cause of my taking medicine to cure an illness; the *content* of my thought of the medicine, recognized as a good, or

ural change has a *terminus* toward which it tends, and which it will attain unless prevented. The eye, when acted upon by light, will produce the intentional species of color unless it is clouded or otherwise incapacitated. Aristotelian physiology classifies the operations of the soul according to their ends, and sets up a hierarchy of ends from the specific to the general: digestion, the natural *terminus* of the stomach's activity, serves nutrition, which in turn serves the overall aim of all vital operations, the material conservation of the organization and the perpetuation of the species. Like Gassendi, Suárez and the other Aristotelians whose works I have studied believed that natural purposes were laid open to view. The role of the eye in seeing, the role of the sexual organs in generation, are not matters to be discovered only by deep theoretical penetration into nature's secrets; they are among the phenomena which the philosopher is called upon to explain.

Nevertheless, in contrast with Aristotle himself, the Christian inheritors of Aristotle tended to assimilate natural ends to intentions—those of the Creator.[35] The natural world reveals itself to have purposes because in it we can read the designs of God, and thus his intelligence and foresight. In so doing, the Aristotelians could answer a number of nagging questions about the final cause. How, for example, can it be a cause if it does not yet exist? How can plants and the inferior species of animal, devoid of cognition, be made to follow ends they are incapable of recognizing? The answer is that, to use Thomas's often-cited analogy, they are directed to them as an arrow is directed to its target by the archer. The wisdom they exhibit is not theirs but God's. For higher animals the allocation of foresight is less clear. *Bruta* have the *sensus communis,* and thus are capable of a certain degree of abstraction; they have memory, imagination, and even a kind of judgment that allows them, in Suárez's words, to apprehend the good, but only "materially," not "formally." Sheep fleeing a wolf do not recognize the wolf under the concept 'harmful thing'; they see the wolf, and "by a natural instinct judge it to be worth wanting, seeking, or fleeing [*naturali instinctu judicant sibi esse appetendum, prosequendum, vel fugiendum*]."[36] Nor would we if we ran away without thinking.

(in Descartes' terminology) my thought considered "formally," is the final cause of my taking the medicine.

[35] See Des Chene 1996, c. 6 and references therein for discussion of this point; see also Osler 1996, 2000, forthcoming, and Menn forthcoming.

[36] Suárez *Disp.* 23§10no14, *Opera* 25:889. In the terms of n. 34, the perception in the sheep's soul that impels it to flee the wolf is only the efficient cause, not the final cause, of the sheep's flight. We may even agree that the sheep has a perceptual concept of wolves, but that concept is associated with harm only by way of God's intentions in instituting certain relations among the perceptions, feelings, and actions (or, to put it anachronistically, the "motor programs") of

Ascriptions of ends to nonrational or "natural" agents and their actions rests on our conceiving them as designed by God. In that respect Descartes differs very little from the Aristotelians. The difference lies, rather, in the interpretation of teleological explanations, and their admissibility in natural philosophy.

Descartes treats any ascription of tendencies to natural agents in terms of deliberate intentional action. "That it appears that I took my idea of heaviness partly from the idea I had of the mind rests principally on the fact that I believed heaviness carried bodies toward the center of the earth, as if it contained some recognition [of the center] within itself" (6 *Resp.*, no. 10, AT 7:442). To have a tendency—toward the center of the earth, toward nourishing oneself, toward pursuing edibles—a thing must "recognize" it, and wish to approach the thing it tends to. But recognition is a mode of thought, and every mode of thought inheres in a thinking substance, a mind.[37] Hence only thinking substances have tendencies. Functions, because they imply a distinction between the way a thing actually does operate, and the way it should, and also that the 'should' is intrinsic to the thing, entail tendencies. To say that the function of the heart is to heat the blood, if by that one meant that the heart has, of itself, a tendency to do so would be like treating heaviness as a genuine quality of things. It would be to treat the heart as if it had the power to recognize the act of heating and the will to perform it.

In a world of tendencies, there is a genuine distinction to be made between "natural" and "violent" changes. Natural changes are those in accordance with a thing's tendencies; violent changes are those that frustrate them. To lift a stone upward is to perform a violent act. So too the birth of monsters, in which the natural tendencies of the seed have somehow been perverted, is violent. Tendencies supply a *norm*. They enable us to distinguish what actually happens from what ought to happen, the normal from the monstrous, health from disease.

There is no basis in Cartesian physics for ascribing tendencies.[38] That is one moral to be drawn from the analogy between living things and clocks:

---

the sheep. God designed the sheep to take care of itself, but the sheep does not *intend* to take care of itself any more than the heart intends to heat the blood.

[37] Having a mind, moreover, is an all-or-nothing property. In Aristotelian terms, there is no remission of substantial form: Socrates is all man or no man at all.

[38] Descartes does, of course, use the term 'tendency to move.' Light is said to consist in the tendency to move of second-element particles. But tendency in this sense is dispositional, not functional: it is a brief way of saying that a body *would* persist in its present state if not acted upon by others, and in particular that it would maintain the instantaneous direction of its motion. That it would do so follows from the first and second law of motion, and thus ultimately from the immutability of the divine will—not, it should be noted, from any particular volition. The laws are part of God's design only in the attenuated sense that, having willed that there should

It might be said here that [people whose senses are disrupted by disease] err because their nature is corrupted; but this does not remove the difficulty, because a sick man is no less a creature of God than a healthy man; and it seems no less repugnant [to God's benevolence] that he should have from God a nature prone to error than that the other should. Just as a watch constructed from wheels and weights no less precisely observes the laws of nature when it is badly made and does not correctly indicate the hours than when it satisfies every desire of its maker; so too if I consider the body of man as a certain machine [. . .] I easily recognize it to be equally natural for the body if, for example, being hydropic, it suffers dryness in the throat, which ordinarily implies thirst to the mind, and if its nerves and other parts are so disposed that it drinks something that aggravates the disease, as it is for the body, when there is no fault in it, to be moved by a similar dryness in the throat to drink something useful. (*Med.* 6, AT 7:84–85; cf. *L'Homme*, AT 11:202)

Just as the conception of the use of a clock provides the standard by which to judge that it is not working well, so too the health or sickness of the body are judged by a standard derived from a conception of the use of the body. That use is to conserve the union:

I observe finally, that since each of the motions that occur in the part of the brain which immediately affects the mind causes [*infert*] just one sensation [*sensum*], nothing better could be thought up than if the motion should cause just that sensation which, of all that can be caused, most and most frequently leads to the conservation of man's health. (87)

Sickness and health, then, are, as Descartes puts it, "extrinsic denominations," in just the way that signifying the time is extrinsic to the bits of matter that make up a clock.[39] Nature itself—not, it should be noted, in the sense of

---

be a world of extended substances, and a certainty quantity of motion in that world, it is a consequence of *that* volition that the actual motions in the world should be in accordance with the laws. (It may be also that the form of the laws, to the extent that they require geometry for their statement and application, will depend on God's will if geometry is included among the eternal truths created by God. Even the first law would be instantiated differently in a world with different congruence relations.)

[39] To "denominate" a thing is to designate it under an identifying description, to refer to it other than by an indexical or proper name. To denominate it intrinsically is to do so by a description that refers to a real entity inhering in the thing, or among those of which it is composed ('real' here means: belonging to it independently of our manner of conceiving it). To denominate it extrinsically is to do so by a description that refers to an entity to which the thing is related. 'Radium' is intrinsic (radium does have the power to produce rays), 'curium' is extrinsic.

"God himself, or the coordination of created things instituted by God" (80), but in the sense of the whole assemblage of material things governed by the laws of motion—contains no clocks, but only things that may or may not serve to tell the time, and no eyes, if by that one means things suitable to see with, but only things that may or may not serve the sense of vision.

In the *Traité de l'Homme,* and even more so the *Description,* that theological setting, which alone justifies the language of function and service, is kept in the background. The *Traité* substitutes for the language of optimally instituted relations between movements and sensations the language of imitation, even where the language of institution would in order.

> As for the disposition of the small fibers that compose the substance of the brain, it is either acquired or natural [. . . .] In order that I should tell you what the natural [disposition] consists in, know that God in forming the small fibers has so disposed them that the passages he has left between them may conduct the spirits moved by a particular action toward all the nerves they ought to go to so as to cause the same movements in this machine, as those to which a like action would incite us when we follow the instincts of our nature. (*L'Homme,* AT 11:192)

The norm is set not by the end of conservation but by that of imitating the acts we are accustomed to perform by our nature. Only at one remove does the conservation of the machine enter into its description.

Descartes needn't have changed his mind between writing *L'Homme* and writing the *Meditations. L'Homme* was originally supposed to include parts on the soul and the union. Without having presented his theory of the union, he was not in a position to make the instituted relation between bodily movements and sensations a standard by which to judge the dispositions of the body. The fable of imitation supplies an intention that can be specified independently of the union.

That intention shares, moreover, with the intention that the body be fitted as well as possible to conserve the union, the feature of having as its object the *whole* body. It can therefore take the place of the other in providing a principle of unity for the machine. Let us say that a thing has *intentional* unity just in case it is the object of a single intention by which it is directed to one end, whether that end is imitation or conservation. Intentional unity is in one respect narrower than functional unity, and in another broader. It is narrower to the extent that things may have ends independently of their being intended to serve those ends. We have seen that Descartes denies that natural things do have ends except by way of intentions; but there is no need for us to prejudge the issue.

Intentional unity is broader than functional because certain intentions, those by which one thing is made to *signify* another, require nothing of the representation except that it exist.[40] At the outset of *Le Monde,* Descartes advises the reader that "there can be a difference between the sensation that we have [of light], that is, between the idea that is formed in our imagination by the intermediary of our eyes, and what it is in the objects that produces this sensation in us." (*Le Monde* 1, AT 11:3). The quality of the objects that produces the sensation of light, like words "which signify only through human institution," could well be a sign "established" by Nature to do so (4), and thus need bear no resemblance to the sensation it gives rise to. The institution or establishment of signs may be a more complicated affair than an intention—for the nonce, perhaps, and not habitual—that one thing should signify another. But in either case no resemblance, indeed no natural relation at all, is necessary to its success. That something has the function of signifying, then, does not of itself entail that the thing has any *particular* disposition or qualities at all.

The intention to make something in *imitation* of something else, on the other hand, will require that the imitating thing resemble the thing imitated in some respect. In *L'Homme* the man-machine is made by God so as to resemble us in outward aspect (oddly enough, Descartes includes not only figure but color, which one would think was quite irrelevant to the purpose of the exercise) and in operation. Descartes includes only those "functions" that "can be imagined to proceed from matter, and to depend only on the disposition of the organs"; but that, as we have seen, is, at least for machines of finite duration, no restriction at all on the *actions* of the machine, so long as they resemble ours. Any finite sequence of actions is consistent with the absence of understanding and will. But clearly not just any machine will do. Clocks, for example, will not. As Fontenelle observed later, clocks do not reproduce themselves. Very few actual machines in Descartes' day were capable even of sustained self-movement (Caus's automata require an influx of water, which they do not provide for themselves, and watches need to have their springs wound or their weights reset).

The conditions under which the machine will resemble us are just that it should have our outward shape and that in any circumstance we are likely to

---

[40] Intentions to signify are correspondingly devoid of implications about the nature of the signifiers. There is, it seems to me, no reason why any collection of things cannot be made to signify something else. Imagine an Arcimboldo who, instead of laboriously gathering together fruits and vegetables into the shape of a human face, simply set a sign in front of a vegetable stand that said *Portrait of Galileo.* The collection is unified under the intention, but there need be no relation among, or intrinsic feature of, its elements to ground the intention (other than the conditions under which the elements of the collection can be thought of at all).

encounter it be so disposed as to act as we would act in that circumstance. It should be geometrically similar to us, and simulate our actions. The intention to build a machine to resemble us is bound by those conditions. Its intentional unity—the unity it has as the *one thing* which is intended to simulate me or you—requires that the mechanisms comprised in it which produce its actions should also have dispositional unity.

Nevertheless the intentional unity of a simulating machine cannot be reduced to dispositional unity. There is, first of all, no guarantee that a machine *will* manage to simulate us even if it is intended to. That remains to be proved. Second, it is not clear that the dispositions of the mechanisms comprised in the machine can be conjoined into a single disposition attributable to the whole. The only description of the action of the whole that comes to mind is that of self-preservation. But self-preservation would seem to re-introduce the norms that a reduction to dispositions was supposed to eliminate. Without them it is not clear that we would know under which conditions the machine could be described as intact. A rusted-out hulk of a clock observes nature's laws no less than a clock still ticking. Like seeds recovered from ancient tombs and germinated thousands of years later, a machine can remain dormant for an indefinite time and yet still be intact, ready to spring to life when acted upon in the right way. Although it might seem quite certain that a machine ground to dust is no longer capable of working, one should remember that in a world where matter is infinitely divisible, no portion of matter is too small to contain all that is needed for the operations characteristic of the human machine.

Functions, one might say, live in the space between intentions and dispositions, a kind of hybrid of the two. The intention to imitate the human body, or to produce a machine which will act to preserve the union of body and soul, is not pure like the intention to signify one thing by another. It contains an admixture of the material. This is evident for the functions formerly attributed to the vegetative soul, since Cartesian mind has no part in them. Growth, for example, is change of quantity. In the Cartesian world, there is no way to change the quantity of a body but by accreting other bodies to it (the food of the machines of *L'Homme,* I should note, since it comes from the world of *Le Monde,* is itself a simulation; in the *Description,* on the other hand, the food of *this* world is configured extended stuff). So the animal-machine, if it is to grow as animals do, will need some means of bringing into itself matter to accrete. Since, moreover, animals, unlike houses, grow in every part, the matter to be accreted to the parts of the machine will have to be divided into bits and distributed. Although you don't need a circulatory system to accomplish that feat, experience tells us that in humans and higher animals there is such a system, and it is reasonable to suppose that it can be made to serve the end of growth.

The senses are more complicated. First of all, the intention to imitate and that of producing the machine best suited to preserve the union diverge. The second must take into account the occasional interposition of thought between the action of sensibles on the sense organs and the responses of the machine. But if, like Descartes, we consider in the body only those sequences in which the soul does not intervene, then the second coincides with the first, except that the assumption of normality, which in the project of imitation is screened off,[41] becomes explicit.

Consider vision. It is a means by which the machine can maneuver well among the things around it, and distinguish edibles from inedibles; it is, moreover, a channel through which those things act on it by way of light rays. Though Descartes does not put it this way, part of the analysis of vision in *L'Homme* and the *Dioptrique* consists in describing the information that can be extracted from the luminous pressures on the body, and in particular on the retina. Light travels in straight lines; light rays do not interfere with one another's passage; bodies reflect light; the texture of their surfaces produces changes in light which the soul apprehends as colors. The eye contains, moreover, a device we can show to have the disposition necessary to focus rays entering through the pupil, and which forms an image on the retina. Experience shows, finally, that humans and some animals act as if they had information about the shapes and sizes of objects around them that we may suppose is conveyed to them by light (the absence of contact rules out touch; that sound may serve the same purpose did not occur to Descartes, even though it was believed that echoes are caused by reflected sound).

The operations of vision, then, must be realized by a mechanism consistent with the phenomena I have just sketched, and which can be shown to help conserve the machine. That mechanism has both the intentional unity conferred on it by its purpose (which, as before, exists only in the divine understanding) and the dispositional unity of a mechanism in the sense of Reuleaux: it is constrained to act in a determinate way when a particular force is applied to its initial segment (namely, the retina). The pressure of light on *this* small

---

[41] To imitate the operations of the human body is as a matter of fact to produce a machine which acts so as to preserve itself; but the aim of self-preservation is implicit, and in a sense incidental. A musician without classical training who undertakes to imitate Bach may as a matter of fact write fugues, but the "fugality" of the pieces she writes is, from the standpoint of her intention, incidental.

part of the retina always results in the opening of *this* valve in the brain and no other, the two being linked by a nerve fiber which, in Descartes' scheme, is treated as if it were a rigid rod.

∞

In Descartes' physiology, the operations of the body, though undoubtedly physical, cannot be completely understood except by referring them to ends. Ends cannot be entirely supplanted by dispositions, even in animals. Reference to them can be deferred by fables of imitation. But if the physiology is to escape the confines of fiction, the role of norms in defining the functions of the body must be acknowledged, and with it that of the rational agent, God, whose intentions in creating animals establishes those norms. The ban on the consideration of ends in natural philosophy must be lifted, even if inferences from dispositions to ends are, as Descartes argues in replying to Gassendi, less certain than inferences from effects to causes.

6.5 Substantial Unity

The introduction of a soul into the machine produces something new, the *union*. The Aristotelian would have it that body and soul are incomplete substances, and the union a complete substance. The incomplete substances are really distinct, but that is no bar to their joining to make a thing that is one in the strong sense, *unum per se*. Even though the soul, alone among corporeal forms, has powers that need no matter to operate, the metaphysics of the human being is in essentials that of any corporeal substance. The union presents no peculiar difficulties of conception. The difficulty for the Aristotelian lies rather in understanding how the soul subsists when it is not joined with the body, and whether its separate existence requires an extraordinary act on the part of God.

In Descartes, that difficulty is alleviated by rescinding from material things all powers of thought. For the Aristotelian, only the rational part of the soul can be argued to require no matter for its operations; the sensitive and vegetative souls are inextricably material. For Descartes *no* mode of thought is a mode of extension, or requires extended substance to exist; even sensation, considered as a mode of thought, must be conceivable apart from matter. The demonstration in *L'Homme* that a machine can simulate not only the vegetative functions but also what in animals we call sensing, feeling, imagining, and remembering allows Descartes to deny that the human soul has either a vegetative or a sensitive part. But a soul with only a rational part would, on the basis of arguments then used to prove the immateriality of the human soul, not depend on matter at all, a position tantamount to denying that it is the form of the body, and therefore contrary to doctrines made matters of Catholic faith at the Lateran Council of 1514.[42] Yet that, it seems, is just what the Cartesian soul, stripped of vegetative and sensitive powers, must be.

Yet a Cartesian soul not only reasons and wills but senses and feels. Descartes does not dwell on the condition of separated souls. It is not entirely clear whether a separated Cartesian soul can have sensations and feelings *in actu*. But it runs contrary to his characterization of the difference between thinking and extended substance to suppose that the soul could be deprived of the *power* to sense and feel by its departure from the body. His demonstrations of the union seem to depend on its being in the nature of the soul to take on certain kinds of mode on the occasion of various events in the body. But if that were true only when the soul was joined with the body, the soul's nature would be altered by union—a result not easy to square with the passivity of matter, or with the claim, put forward in the *Notæ in programma*, that even sensory ideas are innate.

It is not my aim to examine these issues in detail.[43] The topic here is the unity of the body. Is the body, when joined with the soul, genuinely one substance? Descartes' last attempt to explain transubstantiation in a letter to Mesland suggests that the answer is yes. But I am not sure that he had a settled view on the matter. We have seen that Descartes' official view rules out appeal to ends altogether from natural philosophy: to intrinsic ends, because there are none; to extrinsic ends, because the will of the Creator is opaque to human understanding. The question here is whether the ban is lifted for the human body when it is joined with the soul, and thus whether in the union it can be said to have intentional unity, not just *sub rosa,* but according to Descartes himself. Does the brief mention in the *Passions* of the mutual dependence of the parts of the body apply to all animal-machines or only to the human body in the union?

[42] See Des Chene 1997 for a comparison of arguments in Suárez and Descartes. On the "propositions to be held by faith" among Catholic authors, see Des Chene 2000, c. 2.

[43] See Hoffman 1987, which includes a survey of the literature to that point, Voss 1994, Chappell 1994, and Bitbol-Hespériès 1994.

1. The Character of the Union

First a few points regarding the union. Regius, at a time when he was still willing to accept the tutelage of Descartes, proposed the thesis that the human being is an *ens per accidens*.[44] In the terminology of Aristotelianism, that implied
that the union of soul and body was no stronger than that between substance
and accident. The implication would be either that the soul is an accident of
the body, and not its substantial form, or else that the soul is in the body as a
pilot in his ship, or (more pertinently) as an angel would be were it to exert
power over a body. Descartes advised him to say instead that the human being is an *ens per se,* just as the Aristotelians do. Not merely for prudence' sake,
but on the grounds that the soul is, uniquely, a *substantial* form, and with the
body produces something like an Aristotelian complete substance (To Regius,
Jan 1642, AT 3:503, 508). The existence of the union "each person experiences in himself without philosophizing," but the manner of the union we
can conceive only with difficulty, or perhaps even not at all, because to conceive of it we must conceive of the mind and body as *one,* when everything
in our ideas of them entails that they are *two* (To Élisabeth, 28 Jun 1643, AT
3:693–694; cf. *Med.* 6, AT 7:81).

Like Hoffman, then, I understand Descartes to be in agreement with the
Aristotelians on the character of the union, with one necessary correction.
Hoffman writes of form "inhering" in matter.[45] The term 'inherence' was by
Aristotelians reserved for the relation between accidents and substances; substantial form is said to compose or be a component of a complete substance.
This is a basic distinction, found in the first lines of Aristotle's *Categories,* between what is said to be *in* a thing (accidents) and what is said to be *of* it (genus
and species). The corresponding distinction *in re* is between inherence and
composition or union (of matter and form). Because the logic now ordinarily used has no place for such a distinction, philosophers tend to ignore it, or
misread it as a distinction between essential and nonessential properties.[46]

---

[44] On Regius's theses and the controversy that arose around them, see the introduction to
Verbeek 1993, and for further details Rodis-Lewis 1959, Verbeek 1988, Verbeek 1992. The correspondence between Descartes and Regius, and extracts from related documents, are in AT 3.
Prudence was indeed in order, since Regius's denial of substantial forms came under heavy fire
from Gisbertus Voetius, rector of the Reformed Church in Utrecht, who associated Regius's
views with those of bad characters like Basso, Sennert, and Gorlæus (see, for example, AT
3:515, 604).

[45] Hoffman 1987:350–351, and n25. There are similar troubles in Radner 1971. See Moyal
1991, 3:281–282.

[46] I don't think we need to saddle Descartes, as Hoffman does, with the view that "a substance can be a quality." It is true that 'needs nothing to exist but God'—the definition of created substance (*PP* 1§51, AT 8/1:24)—does not *entail* 'cannot inhere in another.' Descartes'
word 'need' (*indigere* in Latin, *avoir besoin* in French) is vague (some critics noted, for example,

But Descartes cannot *quite* maintain the Aristotelian position without qualification. Cartesian soul and body, unlike substantial form and prime matter, are not *incomplete* substances. Descartes identifies 'substance' with 'self-subsistent entity,' self-subsistent, that is, but for requiring the concurrence of God. Incomplete substance, if 'incomplete' means 'not capable of subsisting by itself,' is a contradiction, like a mode not inhering in any substance. Neither can exist even by the absolute power of God (4 *Resp.*, AT 7:222, 434–435). He has only himself to blame if we now find it difficult to understand his view. He was among those who did the most to destroy its logical basis.[47]

Nevertheless, the view is the only one worked out in any detail by Descartes, and the only one that does not court the danger of making the soul an accident of the body, or else a *forma assistens*. Descartes' insistence that thought is sufficient to constitute a substance, and the complete mutual exclusion of modes of thought and modes of extension, would indeed show that the soul is not an accident, but nothing rescues him from the accusation that the hu-

---

that the body needs food to survive). But I am inclined to think that Descartes, rather than do yet more violence to the logic of substance than he had already done by throwing out incomplete substances, was here trying to make the substantial union of complete substances, neither of them moonlighting as a quality, a coherent conception. For Descartes to admit "substantial qualities" would also void his own criticisms of "real accidents," i.e., accidents capable of subsisting with no subject of inherence, like the qualities of the host after transubstantiation.

Hoffman quotes one version of the argument (AT 7:434) but he takes Descartes to conclude from it that heaviness, though a quality, can be a substance (even if in fact it isn't). I think the intended conclusion is that the Aristotelian conception is *incoherent*, not just empirically false. Descartes is not seriously entertaining the thought that possibly heaviness is a quality, he is engaging in a *reductio* of the Thomist doctrine that in transubstantiation the accidents of the Host miraculously subsist without a subject, consistently with his response to Arnauld (AT 7:252–3). Descartes grants that the expression 'una substantia alteri substantiæ accidit' has a coherent sense, but in the very same breath he denies that if a substance *A* 'accidit' another substance *B*, then *A* is an *accident* of *B*.

His example (435) is taken from the rather embarrassing Aristotelian category of *habitus* ('being clothed' and the like, not to be confused with the *habitus* which is a species of quality). Citing Thomas, Toletus classifies *habitus* among the categories that are "predicated of first substances extrinsically, because they do not inhere in them." Like Descartes he takes the predicate in question to be *esse vestitum*, and the *vestis* to be a substance (Toletus *In Log.*, Cat. q25, *Opera* 190b). So if one said 'this bread is hot' and *also* said 'this hotness can subsist without the bread by the absolute power of God,' one would, like it or not, be treating 'hot' not as a genuine accident like *calor*, but as a pseudo-accident like *vestis*.

[47] Des Chene 1996:79n43. The Aristotelian categories created difficulties of their own, not least in understanding what sort of being to attribute to the union. It is not a relation, since 'relation' is an accident; it is not the complete substance itself, since, for example, it is not a subject of predication, nor is it a member of any species or genus. Generally it was classified as a mode, along with other puzzling entities like existences and inherences (see Suárez *Disp.* 7§1n016–22, *Opera* 25:255–258). The relation between Descartes' use of the term *modus* and Aristotelian uses merits further study, as do the numerous sections of *cursus* devoted to the real, formal, modal, and rational distinctions.

[143]

man soul is a *forma assistens,* operating its body as an angel would. Causal relations certainly do not. On the contrary, the definition of soul as *res cogitans,* together with Descartes' emphasis on the action of the soul on the body through the pineal gland, could not help but elicit, when combined with the claim that it is a complete substance, the accusation.[48] In what follows, therefore, I take the union to be substantial union, and the human being to be analogous to an Aristotelian composite of matter and form.

2. Substantial Unity of the Body

The question now is: does the body itself—or rather, not to prejudge the question, does a certain collection of solids and fluids—become one thing in the union?[49] You might think the answer was obviously yes. The soul is one thing; how can it be "substantially united" with several? Wouldn't the result be a Hydra, only much worse, since the human machine has so many parts (perhaps even infinitely many)?[50] A logical monster indeed. Add to this Descartes' position that the soul is "truly joined with the whole body," which implies, of course, that there *is* a whole for it to be joined with.[51] In fact he immediately adds:

> and it cannot properly be said that it is in one of its parts to the exclusion of others, because it is one, and in a certain fashion indivisible, on account of the disposition of its organs, which are so related to one another that when one is removed, it renders the whole body defective; and because [the soul] is of a nature that has no relation with extension, nor to dimensions or other properties of the matter of which the body is composed, but only to the whole assemblage of its organs.[52]

[48] See Arnauld, 4 *Obj.,* AT 7:203. The charge was echoed in this century by Maritain, who accused Descartes of "angelism" (Rodis-Lewis 1971:543–544). See also Voss 1994.

[49] A succinct statement of the issues, and a response to them on Descartes' behalf, is given by Rodis-Lewis (Rodis-Lewis 1990:28–30). See also Rodis-Lewis 1950:60–67, 229–230.

[50] The circulatory system would seem to provide actual instances of the situation envisaged in the *Principles* (PP 2§34–35, AT 8/1:59–60), where a circular stream of particles is confined between eccentric solid rings. In that situation it is necessary that "all its imaginable particles, which are truly innumerable, are moved by one another bit by bit."

[51] "L'âme est véritablement jointe à tout le corps" (*PA* §30, AT 11:351). Of course one can turn this into a tautology, if *tout le corps* simply means 'whatever parts of matter the soul is joined with.' But Descartes, even if he is echoing a traditional claim, does not seem to be treating it as true by definition.

[52] "On ne peut pas proprement dire qu'elle soit en quelcune de ses parties, à l'exclusion des autres, à cause qu'il est un, & en quelque façon indivisible, à raison de la disposition de ses organes, qui se raportent tellement tous l'un à l'autre, cela rend tout le corps defectueux; & à cause qu'elle est d'une nature qui n'a aucun raport à l'estendue, ny aux dimensions, ou autres proprietez de la matiere dont le corps est composé, mais seulement à tout l'assemblage des organes" (*PA* §30, AT 11:351).

In a letter, soon to become notorious, that Descartes sent to the Jesuit Denis Mesland in 1645, he begins a new explanation of transubstantiation with some observations on the body:

> First of all, I consider what the body of a man is, and I find this word 'body' very equivocal; for when we speak of a body in general, we mean a determinate part of matter, and overall of the quantity of which the universe is composed, such that one cannot remove the smallest amount from this quantity without our judging that it is less, and no longer whole [*entier*]; nor can one change any particle of this matter, without our thinking that the body is no longer entirely the same, or *idem numero*. But when we speak of the body of a man, we mean not a determinate part of matter, nor one that has a determinate size, but only all the matter that is together united with the soul of this man; so that, although this matter changes, and its quantity grows or diminishes, we always believe that it is the same body, *idem numero,* while it remains joined and substantially united to the same soul; and we believe that the body is whole [*tout entier*] while it has in itself all the dispositions required to conserve the union. (To Mesland 9 Feb 1645, AT 4:166)

Needless to say, this doesn't answer all the questions. Time now to begin asking them:

(i) The body is "indivisible," but only "en quelque façon." There is a causal dependence among its parts. To say that the removal of one renders the whole ("tout le corps") defective presupposes that one already knows what the whole is, unless 'tout le corps' is taken distributively, as in 'president of all the people.' We begin with the Hydra, and meld all its heads together using the criterion of mutual dependence. But we must have known already which parts of matter the soul is united to. What else would make them "parts of the body" except union? But then the indivisibility-after-a-fashion of the body yields no argument for the conclusion that the soul is not united to some part rather than another, because everything that counts as a part does so by virtue of being united. As the letter to Mesland says, *this* human body is just "all the matter that is united with" *this* soul.

(ii) The soul has "no relation" (*aucun raport*) with any extended substance or with its dimensions. Yet it does have a relation to the "whole assemblage." We know what 'no relation' means: aside from the modes common to all created substances, existence and duration and the like, no mode of extended substance is a mode of thinking substance and conversely. Certainly the size of the body, or the relative sizes of its parts, could have no relation—here roughly in the sense of "common measure"—with the soul; the letter to Mesland goes on to note that the body changes its dimensions a great deal as we grow up. How then could putting a bunch of bodies together make a *rapport* possible, where there could not have been one before? The soul, Descartes says at the end of the article of the *Passions* I have cited, "separates entirely from the body when the assemblage of its organs is dissolved." (The alternative, I suppose, is that it might withdraw from some but not all of the body, as color from a terrified person's face.) True enough, one might say, but that supposes a relation, a mutual fitness that Descartes elsewhere seems to deny:

> it can be objected that it is not accidental [accidentarium] to the human body, that it should be joined with the soul, but rather this is its very nature [ipsissimam eius naturam]; because when the body has all the dispositions required to receive the soul, and without which it is not properly a human body, it cannot happen except by a miracle that the soul should not unite with it; so also it is not accidental to the soul, that it should be joined with the body, but only accidental to it after death, that it is separated from the body. All these things should not be denied altogether, lest the Theologians be again offended; nevertheless one should respond, these things can be said to be accidental for the reason that, considering the body alone, we perceive nothing in it on account of which [propter quod] it should desire to be united with the soul; and nothing in the soul, on account of which it ought to be united with the body, and so I said a little bit later, it is in a certain way accidental, but not absolutely accidental. (Descartes to Regius, mid-December 1641, AT 3:460−461)[53]

---

[53] See the remarks on this passage at the end of Hoffman 1987. The cluster of terms formed off the stem *accid-* (*accidens, accidere, accidentalis, accidentarius* and so forth), and phrases like *per accidens,* is, like the cluster from *es-* (*esse, essentia, ens, essentialis, entitas,* etc.), utterly important to understanding philosophical positions in early modern philosophy and yet very confusing to readers now, especially since the vernaculars tend to take over only some of the terms in each cluster. 'Accidentaire' is possible in French, 'accidentary' in English, but neither is used, and so I had to translate *accidentarius* with 'accidental,' thereby blurring the distinction with the already polysemous 'accidentalis.' *Accidentarius* seems to be used as the adjective corresponding to *per accidens:* if my intention to go to the store is the cause *per accidens* of my meeting you (there is nothing in the nature of an intention to go to the store that grounds a relation between it and meeting you), then it is *accidentarius* to my intention that it should have caused me to meet you, even though, in a certain sense, it did (e.g., it was a necessary condition of my going to the store at all).

Though it is possible to doubt that Descartes is expressing his own view with entire candor, since he is advising Regius in his dealings with the professors, still it is also true that the tone of the correspondence (of which we have little from Regius) is one in which Descartes is casting Regius as his spokesman, to pronounce publicly what Descartes believes privately, even if in guarded terms. In any case the message is clear: nothing in the nature of the soul provides grounds (*propter quod* can denote both causal and teleological grounds) that it should be united with the body, and so too for the body.

(iii) From the very fact that, as Descartes reminds Mesland, the parts of the body are constantly changing (every part of it is fluid, but some more than others, see §2.3), it follows that some bits of it can be removed without affecting others. But the parts in question are "organs": which parts are they? Perhaps just those whose removal does render all the rest defective. But the spleen, a kidney, a hand, an eye can be lost without making other organs defective. Either the list of organs is shorter than common sense and the anatomist would have it, or else 'organ' can be defined independently of mutual dependence. How it is to be defined without appealing to ends is not clear, for reasons given already. The shape of a part does not make it an organ, or the events that go on in it: what would lead us to choose some shapes and not others?

(iv) The dissolution of the body is the end of the union. We have seen that from the standpoint of physical processes alone, the conservation of the body (and so its defective states too) are not easily defined. "A horse whose hoof is cracked is broken" (Rodis-Lewis 1990:30). But what is broken about it? Part of it is no longer whole in the strict physical sense. But not every solution of continuity is breakage, nor of course is every defect a rupture of bodily parts. The horse no longer runs well, perhaps. But then we are applying a notion of normality—and merely statistical normality will not do.

The pieces of the puzzle begin to fit together if one keeps a steady eye on death. Death typically requires bodily change, even if it is by definition the destruction of the union and nothing else. Yet how should it be that *any* bodily change should bring it about that the soul ceases to be united with that body (or any body)? Some do: let's call them "dissolutions of the body." They differ from mere fractures and bruises because after they occur, we find that the body ceases to exhibit any of the behavior from which the presence of a soul can be inferred. No more in humans than in animals do we perceive even the bodily processes that are called imagining and sensing (To Gibieuf, 19 Jan 1642, AT 3:479), and still less thinking and willing.

There is nothing in the nature of a human body that would ground the assertion that it should be joined with a soul.[54] Adding to the bare nature of *res extensa* a blueprint of the human machine does not alter the indifference of matter to mind in the least. Similarly the nature of human soul, considered only as *res cogitans,* contains nothing that would ground the assertion that it should be joined with a body. If there were, the argument for the real distinction between mind and body would begin to falter, because it requires that (in the words of the Fourth *Replies*) mind should have none of the "forms or attributes, from which we recognize that body is a substance" (AT 7:223). In particular, it seems to me that the mind cannot, considered *only as mind,* have what Suárez calls a "transcendental habitude" toward body.[55]

And yet there is such a thing as death. There are conditions under which the union can no longer naturally exist. (Whether there are conditions under which it must exist, or under which "it cannot happen except by a miracle that the soul should not join" with the body, is another matter.) Those conditions, I should note, pertain entirely to the body. There is, so far as I can tell, no condition of the *soul* that would cause the union to cease. Merely wanting to die is obviously insufficient. We must suppose that the conditions of dissolution are grounded not in the nature of body, nor in that of soul, nor in the laws of nature, but in something else.

The *Meditations* tell us that the "corruption" of the nature of the person who wants to drink water even when it is harmful is "only an extrinsic denomination" with respect to the body. But "with respect to the composite, or the mind joined to such a body, it is not a pure denomination, but a true error of nature" (AT 7:85).[56] It is an error with respect to the composite because God has instituted between the motions in the brain and the modes of thought they give rise to that relation which, when a certain motion occurs

[54] Descartes says 'that it should desire [*desideret*]' a soul, perhaps recalling the slogan 'matter wants form' [*materia formam appetit*] which is often made the topic of a question in *Physics* commentaries (see, for example, Toletus *In Phys.* 1c7q17, *Opera* 4:37vb–38ra). But the point is not the relatively trivial one that matter has no desires. It is that in the nature even of a fully formed human machine, there are no grounds from which one may infer that it will, or ought, to be joined with a soul.

[55] Suárez *De An.* 1c3n020; *Opera* 3:490. See Des Chene 2000, Chapter 5.

[56] Paraphrased without the technical term *denominatio,* Descartes' claim is that with respect to the body 'corruption,' as a name for the body's state, refers not to any property of that state, but to the diminished usefulness of the body, in that state, from the standpoint of the soul; on the other hand, the breakdown of the relations instituted by God between body and soul in the union, or their failure to fulfill their end, *is* a genuine "error of nature." See Gueroult 1968, c17§2, §4–5. The "nature" referred to by Descartes in the passage quoted is what Gueroult calls

as a result of a particular condition of the body, will give rise to sensations and passions that incline the will to actions most conducive to the continued existence of the body in a condition that will normally sustain the union (*Med.* 6, AT 7:87). In a person whose sensations lead to actions endangering the union, the error is not that the instituted relations have been altered, but that the antecedent causes of some motions in the brain are not what they normally are—not, so to speak, what God anticipated in instituting the relations.

The extreme case is when the body is so badly damaged that its motions can no longer serve the end of guiding the soul in preserving it. When the soul is joined with the body, it feels joy, "because it cannot be believed that the soul was put into the body unless the body was well disposed, and when it is thus well disposed, this naturally gives us joy" (To Chanut 1 Feb 1647, AT 4:604). Joy, we learn from the *Passions,* is excited by "the consideration of a present good" (AT 11:376). The body is, precisely, present to the soul in the union, and its being well-disposed is, one must suppose, represented to the soul as good with regard to the soul itself (cf. 11:374). But how is it good with regard to the soul, if "there is nothing in the soul on account of which it ought to be united" with the body? The nature of the soul, considered as *res cogitans,* includes no *telos,* no end that might be achieved with the aid of a body.

It is only with respect to itself considered as *already united* with the body that the body is a good for the soul; the union itself is a brute fact, upon which all other judgments of benefit and harm are predicated. Similarly it is only with respect to the union that the dissolution of the body—to the point where its sensations (or the corporeal ideas of *L'Homme*) no longer serve the soul in acting so as to maintain it—is a "corruption," both in the sense of something bad and in the Aristotelian sense of the destruction of a complete substance. But it is a bad thing that ceases to be bad as soon as it happens, because the termination of the union is the termination of that upon which the judgment of badness was predicated.

An analogy will help illuminate the point. A partnership is a relation among individuals instituted to serve certain ends. We may suppose that a certain amity, or cooperation, among the partners is necessary to the continuation of the partnership. With respect to the partnership, a decrease in cooperation, or a cause of dissension, is bad because it endangers the partnership. But if the decrease is so severe that the partnership is terminated, then cooperation no longer is either good or bad, since judgments of its goodness or badness were predicated upon the existence of the partnership.

"my nature in the broad sense" (ibid. 2:160). Gueroult's exhaustive analysis of the sixth *Meditation* is essential to understanding the teleological constitution of the body.

Now the partnership of body and soul is not, as I said, instituted to serve any ends (or at least there is nothing in the nature of soul alone to ground the supposition that it does; God may have his purposes in joining human souls with body, but that is none of our business except as he chooses to inform us). So the goodness or badness of the condition of the body is *simply* with respect to the union, whereas the goodness or badness of cooperation in a partnership might be regarded as an indirect means to the ends for which the partnership was instituted.

The very notion of dissolution rests on that of the union. As Gueroult writes, "It is the union with a soul that *converts* the relation, in itself purely mechanical, of the parts assembled into this whole into a *teleological relation* with respect to the whole body, and it is by the union that the corporeal organism succeeds in acquiring a genuine functional indivisibility" (Gueroult 1968, 2:185). If the body is a whole only in the union, then its ceasing to be a whole is only with respect to the union. The body considered apart from the union cannot die, no more than the soul, but for a different reason. The soul is immortal because it is not extended, and thus indivisible; the body, considered simply as *res extensa,* is indestructible because no configuration is more natural to its matter than any other.

We have seen that it is not easy to spell out the "mechanical relation" that obtains among the parts of a machine considered simply as a disposition of *res extensæ.* Descartes' suggestion, included almost as an afterthought, that there is, in human *and* in animal machines, mutual dependence among the organs requires us to have a "mechanical" notion of defectiveness and of organ. But at most we can say that we sometimes notice that the ablation of a part (and especially the kind of part known to anatomists as an "organ") causes the rest to operate differently than before.[57] That is very far from any notion of unity—or rather it will take centuries to make the underlying causal relations precise enough to yield a scientific notion that improves much on common sense.

Call the observed covariation of the operations of its parts the "empirical unity" of the machine. It is neither necessary nor sufficient for the union. That it is not sufficient is clear enough: no degree of complexity, no degree of fineness and differentiation in operation, could entail that a machine should be joined with a soul, since, as I have said, no mode of extension, complicated or not, can do so. That it is not necessary, at least with respect to the absolute

---

[57] "But if this end is not really immanent in the machine, there is no defect, whatever may happen to it. I simply observe that the absence of one of its elements gives a different convergence of movements that the one I habitually observe" (Gueroult 1968, c17§6, 2:184).

power of God, is clear from Descartes' explanation of transubstantiation in letters to Mesland from 1645 and 1646. The body, he writes,

> remains always numerically the same as long as it is united with the same soul. [. . .] Whatever matter it may be, and whatever quantity or figure it may have, provided that it is united with the same rational soul, we always take it for the body of the same man, and for the entire body, if it has no need of being accompanied by other matter to remain joined with this soul. (To Mesland 9 Feb 1645, AT 4:167)

Even the tiniest particle of the Host can be the whole body of Christ: "all the matter, however large or small it may be, that is together informed with the same human soul, is taken for an entire human body" (168). Descartes goes on to say that some may find his explanation shocking, but the reason is that such people think that "all the members [of Christ's body] are there with their same quantity and figure, and numerically the same matter of which they were composed when he ascended into heaven" (169). It was this point that would prove decisive in the hostile reception given to Descartes' explanation when it was finally made public. But for present purposes, Descartes' reasons for hesitation in offering the explanation to Clerselier are irrelevant, since they do not touch the basis of the explanation.

Whichever bits of *res extensa* are united to a soul are "bodies" of that soul in the miraculous circumstances of transubstantiation: *whole* bodies, not bits of bodies. Hence the only *necessary* condition for the unity of the body in the union is, precisely, union. One might well ask: how is it that when my soul is joined with what is, after all, not physically nor perhaps even dispositionally one body, I have just *one* body, not millions? Words like 'miraculous' and 'supernatural' won't make the question disappear. Instead they lead to a further question: what is *natural* about the union of the soul with what we usually call the body, and what would be *unnatural* about its being joined distributively to each of the bits of matter in the collection designated by that name?

Clearly the sense of 'natural' here is not that in which it is just as natural for a body to be dead as alive. Perhaps what is natural is only that God has chosen to unite human souls with bodies that have "empirical" unity (and thus perhaps dispositional unity, to the degree that the one is evidence of the other). Just as the relation between a particular configuration of the pineal gland and the mode of thought we call "perceiving red" is, it would seem, entirely arbitrary, so too the relation of union itself is arbitrarily instituted by God to hold between only some kinds of body and human souls, and not others. The correlation between the empirical unity of living human bodies and

their ensoulment would be an inexplicable *factum,* and since there is no argument—as there is with the laws of motion—that the immutability of
God's will should require him to join each soul with the same sort of body,
physiology and medicine can rely only on the fortunate consilience of God's
choices.

But there may be a more promising alternative. Consider once again the
passage from the sixth *Meditation* that I have mentioned several times already:

> I notice, finally, that since, among each of the motions that occur in the part of
> the brain that immediately affects the mind, just that sensation follows from it,
> nothing better can be devised in this matter than if that sensation should follow
> which, of all that could follow, conduces best and most often to the conserva-
> tion of the health of man. (*Med.* 6, AT 7:87)

For 'conservation of health,' read 'continuation of conditions under which
union is naturally possible.' How will a sensation (taken in the broad sense to
include passions) conduce to the conservation of health? By leading to some
sort of action: eating what is nutritious, refusing what is harmful, sitting in
green glens, and the like.[58]

Thus it is not modes of thought taken one by one whose relation to the
body is instituted by God to be the best suited to conserving health. It is the
whole complex consisting of brain movements (understood in relation to
their typical causes), sensations and passions, and the inclinations of the will
resulting from those passions. In that sense only something very much like the
whole body, empirically and dispositionally one, can rightly be said to be the
object of God's intention.

It follows that in the human body, there is a kind of hierarchy among the
various kinds of unity I have described. Physical unity, especially of the lax
sort, is necessary in certain instances for dispositional unity, which in turn is
necessary if the functions of the union—the operations by which it conserves
itself—are to occur reliably; and those functions themselves are defined only
in relation to the divine institution of the union. Nevertheless, the only real,
the only metaphysical reason to call the body *one thing* is that God has willed
that this collection of parts of matter should be our instrument.

---

[58] "La verdeur d'un bois": To Élisabeth, May or Jun 1645, AT 4:220. According to
L'Homme, green is the most moderate of colors.

Conclusion

Throughout his natural philosophy Descartes was too ambitious by half. Consider what was needed finally to understand the inheritance of characters. You would have to include organic chemistry, the cell theory, the discovery of mitosis, the identification of genes, the analysis of DNA and of the mechanism of replication—each of them a major achievement.[1] A single lifetime, a single mind, even Descartes', were certainly not sufficient to complete the science of life. Not only did Descartes not manage to complete the science of life: in respect to particulars, he failed even to begin it. Within fifty years of his death, most if not all the mechanisms proposed by him were rejected outright, as were the *feu sans lumière* in the heart, the role of the pineal gland in sensation and memory, and most of his embryology, or substantially modified.

1. The Question of Emergence

Nevertheless Cartesian physiology carried the day on one point. The vegetative soul was dismissed. That may seem now like a minor affair, much less noteworthy than the elimination of the sensitive soul. But when the human soul ceased to have, in common with every living thing, a vegetative part, when it ceased to play any role in generation, a link between the intellect and the life of the organism was broken. It is true that Descartes adduces the

---

[1] These are only some of the milestones in biology itself. From other fields one would have to add, for example, crystallography and the analysis of crystalline structures by X-rays to the list, since the techniques of that analysis were essential to the discovery of the double helix. Descartes foresaw (his contemporary Mersenne was more forward-looking, at least in practice) neither the conceptual nor the social revolutions that had to occur to make an understanding of inheritance possible.

concern for the body's condition evoked in us by the passions as evidence of the union; true also that in a letter to Chanut he agrees that the soul, upon being united with the body in the womb, feels joy, and recognizes the body to be good for it. There is not much sign, especially in his later work, of contempt for the body—certainly not if one compares him with Malebranche or Pascal; the body is, by way of the sensations and passions we have when our soul is joined with it, a hindrance to inquiry; but Descartes does not suggest that we should neglect or abuse it. The beast does need to be tamed, but only to the point at which we understand how to use the passions, and to rule them. There is no virtue in going beyond that.

Nevertheless neither the vegetative nor, in a certain way, even the sensitive functions of the soul have any *cognitive* significance. More precisely: the senses and the passions (some of which, like the desire for food or for coition, are brought about by changes in the body formerly referred to the vegetative part of the soul) are indeed a guide to preserving the union. They provide signs by which we can learn to make them serve that end. But those signs are opaque. No more than a cry resembles the injury that evokes it do sensations of color or feelings of dread resemble the physical conditions that, through the instrument of the body, give rise to them.

Moreover, the body for its part has no need of the interpretations of the sensible world invented by the intellect to preserve its functioning. Descartes shows how the cycles of bodily activity—the pulse, the wasting and replenishment of the blood, the consumption and the acquisition of food—can be explained: the body is to be analyzed into systems of mechanisms, and each mechanism into simpler mechanisms, until we arrive at mechanisms whose capacities can be understood in terms of the modes and derived properties of extended things. Self-motion is no obstacle to mechanism.

Descartes, it would seem, succeeded in making mechanism an all-or-nothing affair. With due allowance for anachronism, it is reasonable to think of Aristotelian souls—the human excepted—as something like emergent properties (though they aren't, strictly speaking, properties). They are not reducible to the elementary qualities of heat, cold, wet, and dry, nor to combinations or "temperaments" of those qualities, such as mixtures like flesh and blood were said to have. What in the twentieth century appeared as the "problem of emergence" has its parallel in the Aristotelian tradition as the problem of the eduction of forms: if the forms of plants and animals do not exist potentially in matter, then where do they come from? The standard answer, for higher animals and humans at least, is that they come from the heavens or from God. Descartes does not solve the problem of eduction. He gets rid of it. There are no souls in animals or plants, and thus no eduction of forms.

In the vocabulary of the present day: there are no emergent properties in the world of Descartes.

It is striking that among those major figures who rejected Cartesian mechanism, the tendency was not to resurrect forms, or to argue for something like emergence, but rather to extend Cartesian soul, perhaps in some attenuated version, to everything. Leibniz with his monads, Spinoza with his assertion that everything is indissolubly a mode both of thought and of extension, the Cambridge Platonists with their "plastic natures," agreed, it would seem, that there is no emergence. But their conclusion was not that only in human bodies are there souls, it was that the mental goes all the way down. The *mental,* I should note, and not the mere vitality of the vegetative soul. Life is bound up with perception, even if that perception is, as Leibniz says, obscure and confused in plants and in things we call dead.

2. Machines and Simulation

The appeal of the animal-machine diminished considerably by the end of the seventeenth century. But simulation remained among the tools of scientific explanation, as did questions concerning its import. There are, of course, well-known questions about alternative mechanisms—questions raised by Descartes himself. I have been concerned with other issues. Descartes, in proposing the fiction of the divinely constructed automaton in *L'Homme,* established, perhaps without quite realizing its full implications, an important strategy in science. The fiction of imitation allows him to *postpone*—indefinitely, as it turned out—questions about the origins of living things, both about the origin of species and about the generation of individuals. It allows him, moreover, to *isolate* the organism from its surroundings. *L'Homme* doesn't ask where the automaton's food comes from: it simply attempts to explain how, once the food is ingested, it is turned into blood.

Cutting off the organism from its origins and surroundings is an application to the whole of the analysis of capacities which in §4.2 I discussed only in relation to parts. The engineer dissects the machine into mechanisms, whose actions may then be analyzed, and which may be extracted and used in new machines. Essential to that strategy is to treat the influences coming in and the products going out as given—as issuing from or disappearing into a black box, so to speak.

The role of God in the fiction of imitation is in part to *trivialize* questions of origin and of the environment. Instead of puzzling ourselves about origins, we simply give ourselves, by positing a divine act of creation, the organisms to be studied, the food they eat, and the stimuli that move their senses. Descartes begins the account of generation in the *Description* with the mingled seed of the two parents who, like the God of *L'Homme,* have left the stage before the play begins. Only the womb of the mother remains as a container for the fetus, its role limited to that of occasionally redirecting the flow of fluids. There may be, in the isolation of the fetus from the body of its mother, a rejection or occultation of the feminine. But it is also a means by which to sever links to earlier causes and effects. There is no ancestral memory, nor anticipation of fruits. Only by way of the *present* mechanical properties of the seed can its development be explained. The same holds at every stage of explanation, from the initial outflow of blood from the future heart to the last filigrees of capillary and nerve.

The fiction by which the automata of *L'Homme* are given to us has another function as well. God, using only the matter, which is to say the extended substance, of Descartes' world, *imitates* our bodies as closely as possible. How closely is that? The standard of resemblance cannot be that he reproduces the *nature* of our bodies. Precisely that is in question. The standard is instead their *operations.* Resemblance in operations: but can these be described without begging the question? Descartes needs an austere, one might say a chastened, vocabulary, which even if it sometimes uses the same words as the Aristotelian or the Galenist, cannot take on all their implications.

Consider seeing. Seeing is not, for example, sensing the colors of things. There are, in Descartes' world, no colors; there are only textured or tilted surfaces and rotating particles of light. To see, then, is *to be affected by* light. But not in just any manner: not only the retinas, but the skin of the machine is affected by light. Seeing is being affected by light in such a way as to bring about changes in the course of the animal spirits of the brain, and thereby influence the actions of the machine. It is crucial to Descartes' physiology that the machine be represented as responding appropriately to the world around it. If we do not consider its actions, we have no way to distinguish sensing from other effects of external bodies on the machine.

The fiction of resemblance begins with machines that resemble us as much as an object made of Cartesian matter can. It has our figure, our size, it moves as we do. But *because* it operates as we would in the same circumstance (barring the intervention of the soul), we are to infer that the machine is *indiscernible* from the human body. Although Descartes speaks of "not needing" the vegetative soul or the various powers that Aristotelians ascribed to living things, the argument of *L'Homme* is not simply an application of Ockham's razor. I have already mentioned the need to describe the phenomena in the right way. In Descartes' story of the stranger who, from a world of machines and no animals, comes to ours and concludes that our animals are machines, the point is not merely that he cannot tell the difference, and has therefore no reason to suspect that the sensible qualities and actions of our animals spring from different principles than those of his machines. It is that the stranger understands machines *mechanistically,* and imputes to them no principle of operation not authorized by Cartesian physics. Machines need not be so understood, not even if one builds them oneself. Descartes relies on the easy move from 'this is a machine' to 'this operates solely on mechanistic principles': that move erases the gap between the indiscernibility of sensible qualities and operations and the identity of principles.

3. Norms and Unity

To imitate something, one must know what the original is. In the fiction of resemblance in *L'Homme,* God imitates, it would seem, a *normal healthy* human body, not a monster or a body wracked by disease. In the *Description,* the fiction of resemblance is dropped. Nevertheless the object of description remains a normal healthy body. I have argued that without a norm supplied by nature the choice of the object of imitation or description is arbitrary, or at best governed by what we regard as a good sort of body to have. We would, in other words, treat it as one instrument among others, for whatever aims we choose. The standard of health, and the aims of therapy, would thereby be determined.

In the discussion of unity, the body—insofar as it answers to the divine intention that we be provided with a guide to conserving the union by way of our sensations and passions—has *intentional* unity. The norm required in the description of the body is here supplied: a normal healthy body is one in which the divine intention is fulfilled, so far as this is possible given the limitations of the senses and the mechanisms of action.

One question remains. The body is treated by the soul as good—as being of benefit to it. But it is of benefit only with respect to conserving the union. Otherwise it is unclear what the soul gains, and Descartes insists time and again on the disadvantages of the body when it comes to the exercise of the intellect. The condition of health is desirable only insofar as the continuation of the union is. What Descartes needed, so long as he maintained the divine institution of the relations of body and soul in sensation and passion, was an anthropology. Why is there such a thing as a human soul? What is its place in the order of things? In the *Meditations* we learn that the mind is a mean between God and nothing: but so too is every created thing. We can, of course, resign ourselves, as the Fourth *Meditation* recommends, to our condition. And that, perhaps, is what a Cartesian ought to do. The *Passions,* it is true, recommend *générosité* as the passion by which one may "follow virtue perfectly" (*PA* §153, AT 11:446). But that virtue is strictly of this world. It consists in understanding clearly that only our volitions lie entirely within our power, and in acting firmly on them: in knowing, that is, what we can do, but not why we are such as to have the powers we have. Though Descartes' strictures against the appeal to ends in natural philosophy are less rigorous than his explicit doctrine would lead one to believe, still the question of the end for which we souls pass through this world is one he no doubt thought was best left to faith.

BIBLIOGRAPHY

PRIMARY SOURCES

Aristotle
De An.
Aristotle. *De anima.* Latin translations are from the *versio antiqua* of Moerbeke. See Thomas
    Aquinas *In de An.*
De Gen. et corr.
Aristotle. *De generatione et corruptione.* Greek text and Latin translation. See Coimbra *In de*
    Gen.
De Motu anim.
Aristotle. *De motu animalium.* Trans. and comm. Martha Craven Nussbaum. Princeton,
    N.J.: Princeton University Press, 1978.
Gen. anim.
Aristotle. *De generatione animalium.* In: *Opera* 350–384 (trans. Theodorus Gaza).
Opera
Aristotle. *Aristoteles Latine interpretibus variis.* Ed. Academica Borussica. Berlin: Georgius
    Reimerus, 1831. (The translations are mostly Renaissance translations by Julius Pacius,
    J. C. Scaliger, and J. Argyropoulos, and others.)

Arriaga
Cursus
Arriaga, Rodericus de. *Cursus philosophicus.* Antwerp: Balthasar Moretus, 1632. (Ten sub-
    sequent editions, including four in Paris, through 1669. Facs. microfilm, *Manuscripta:*
    microfilms of rare and out-of-print books: list 84 reel 7.)

Baglivi
Praxis medica
Baglivi, Giorgio. *De praxi medica ad priscam observandi rationem revocanda.* Libri duo. Accedunt

dissertationes novae. Rome: Typis Dominici Antonii Herculis, sumptibus Caesaretti, 1696.

Baillet
Vie
La Vie de Monsieur Des-Cartes. Paris: Daniel Horthemels, 1691. (Facs. repr. New York/ London: Garland, 1987.)

Bartholin
Carmina
Bartholin, Thomas. *Carmina*. Hafniæ: Daniel Paullii, 1669.

Bauhin
Theat. anat.
Baglivi, Giorgio. *Theatrum anatomicum*. Novis figuris aeneis illustratum, et in lucem emissum opera et sumptibus Theodori de Bry. Frankfort: M. Becker, 1605.

Beeckman
Journal
Beeckman, Isaac. *Journal tenu par Isaac Beeckman de 1604 à 1634*. Ed. Cornelis de Waard. La Haye: M. Nijhoff, 1939–1953.

Boerhaave
De usu
Boerhaave, Hermann. *De usu ratiocinii mechanici in medicina oratio*. Leyden: H. Teering, 1709.

Boullier
L'Âme des bêtes
Boullier, David Renaud. *Essai philosophique sur l'âme des bêtes*. Précéde du *Traité des vrais principes qui servent de fondement à la certitude morale*. Paris: Fayard, 1985. (Text from the second edition: Amsterdam: François Changuion, 1737. Orig. publ. 1728.) (*Corpus des œuvres de philosophie en langue française*)

Cardano
De subt.
Cardano, Girolamo. *De subtilitate*. In: *Opera* 3:357–672.
Opera
Cardano, Girolamo. *Opera omnia*. Lyon: Ioannes Antonius Huguetan & Marius Antonius Ravaud, 1663. (Facs. repr. Stuttgart-Bad Cannstatt: Friedrich Frommann, 1966.)

Caus
Forces mouvantes
Caus, Salomon de. *Les raisons des forces mouvantes, avec diverses machines tant utiles que plaisantes. Aus quelles sont adioints plusieurs desseins de grotes et fontaines*. Frankfurt: Jan Norton, 1615. Also Paris: C. Sevestre, 1624. (Fasc. repr. of the 1615 ed.: Amsterdam: Frits Knuf, 1973.) (*Bibliotheca organologica*, 21)

Clauberg

Opera

Clauberg, Johannes. *Opera omnia philosophica; ante quidem separatim, nunc vero conjunctim edita, multis partibus auctiora & emendatiora* [. . .]. Cura Joh. Theod. Schalbruchii. Amsterdam: P. & I. Blaeu, 1691.

Coimbra

In de An.

Commentarii Collegii conimbricensis Societatis Iesv, in tres libros De anima Aristotelis Stagiritae . . . Coloniae [Cologne]: Lazari Zetzneri, 1600. (The author of this commentary is Emmanuel de Goes.)

In de Gen.

Commentarii Collegii Conimbricensis Societatis Iesu, in duos libros de Generatione et corruptione Aristotelis Stagiritæ. Lugdunum [Lyon]: Horatius Cardon, ²1606. (The author of this commentary is Emmanuel de Goes.)

In Parv. nat.

Commentarii Collegii Conimbricensis Societatis Iesu, in libros Aristotelis, qui parva Naturalia appellantur. Lugdunum [Lyon]: ex Officina Iuntarum, 1598.

Descartes

De Hom.

Descartes, René. *De Homine, figuris et Latinate donatus a Florentio Schuyl.* Lugduni Batavorum: Franciscum Moyardum et Petrume Leffen, 1662.

Descrip.

Descartes, René. *La description du corps humain.* In: *Œuv.* AT 11:223–286. (Orig. publ. with the *Traité de l'Homme* under the title *De la formation du fœtus.*)

Diop.

Descartes, René. *La Dioptrique.* In *Œuv.* AT 6:81–227. (Orig. publ. as one of the *Essais* with the *Discours* in 1637.)

Discours

Descartes, René. *Discours de la méthode, pour bien conduire sa raison, et chercher la vérité dans les sciences* [. . .]. In: *Œuv.* Alq. 1:568–650; AT 6:2–78. (Orig. publ. Leyden: chez Jean Maire, 1637.)

Entr. Burman

Descartes, René. *L'Entretien avec Burman.* Ed. and tr. by Jean-Marie Beyssade. Paris: PUF, 1981. (Also in *Œuv.* AT 5:146–179. The conversation took place 16 April 1648, but the manuscript containing it was discovered only in 1895, and published in 1896.)

Excerpta

Descartes, René. *Excerpta anatomica.* In *Œuv.* AT 11:549–634. (Extracts copied for or by Leibniz from manuscripts of Descartes in 1675–1676. First published by Foucher de Careil in *Œuvres inédites de Descartes,* 1859–1860.)

L'Homme

Descartes, René. *L'Homme.* In: *Œuv.* Alq. 1:379–482; AT 11:119–202. (Orig. publ. in Latin translation in Leyden in 1662; the first French edition, by Clerselier, was published in 1664. English trans: T. S. Hall, *Treatise of man.* Cambridge, Mass.: Harvard University Press, 1972.)

Le Monde

Descartes, René. *Le Monde*. In: *Œuv.* Alq. 1:315–377 (with omissions); AT 11:3–118. (Orig. publ. Paris: Jacques Le Gras, 1664.)

Le Monde/L'Homme

Descartes, René. *Le Monde, L'Homme*. Ed. with notes by Annie Bitbol-Hespériès and Jean-Pierre Verdet. Intro. by Annie Bitbol-Hespériès. Paris: Seuil, 1996. (*Sources du savoir*)

Med.

Descartes, René. *Meditationes de prima philosophia*. In: *Œuv.* AT 7:17–90; Alq. 2:177–235.

Météores

Descartes, René. *Les Météores*. In: *Œuv.* AT 6:231–366; extracts in Alq. 2:719–761. (First published with the *Discours* in 1637; Latin translation, 1645.)

Œuv.

AT

Descartes, René. *Œuvres de Descartes*. Ed. Charles Adam & Paul Tannéry. Nouvelle présentation. Paris: Vrin, 1964–1991.

Alq.

Descartes, René. *Œuvres*. Ed. Ferdinand Alquié. Paris: Garnier, 1963–1973.

CSMK

Descartes, René. *The philosophical writings of Descartes*. Ed. and trans. J. Cottingham, R. Stoothoff, D. Murdoch, and Anthony Kenny. Cambridge: Cambridge University Press, 1985 (v. 1–2), 1991 (v. 3).

PA

Descartes, René. *Les passions de l'âme*. Paris: Henry le Gras, Amsterdam: Louis Elzevier, 1649. In: *Œuv.* Alq. 3:941–1103.

Primæ cog.

Descartes, René. *Primæ cogitationes de generatione animalium et nonnulla de saporibus*. In AT 11:505–542. (Orig. publ. in *Opuscula posthuma*, Amsterdam, 1701.)

PP

Descartes, René. *Principia philosophiæ*. In: *Œuv.* Alq. 3:83–525; AT 8/1. (Orig. publ. Amsterdam: Ludovicus Elzevirius, 1644. French translation by Picot: *Les principes de la philosophie*, Paris: Henri le Gras, 1647.)

Descartes & Schoock

Querelle

Descartes, René, & Martin Schoock. *La querelle d'Utrecht*. Paris: Les impressions nouvelles, 1988.

Diderot

Élém. physiol.

Diderot, Denis. *Éléments de physiologie*. Ed., intro., and notes by Jean Mayer. Paris: Marcel Didier, 1964. (The date of composition is uncertain; the text was left unfinished at Diderot's death in 1783: see Mayer's introduction, xii–xv.)

Encyclopédie

Diderot, Denis, ed. *Encyclopédie ou Dictionnaire raisonnée des sciences, des arts et des métiers*. Paris: Briasson/David/Le Breton/Durand, 1751. (Facs. repr. Elmsford, NY: Readex/Pergamon, 1969.)

BIBLIOGRAPHY

Du Fresnoy
De arte
Du Fresnoy, Charles-Alphonse. *L'Art de peinture de C. A. Du Fresnoy.* Fr. trans. Roger de
Piles. par 2. ed. Paris, Chez N. Langlois, 1673. (Includes the *Dialogue sur le coloris* by
Piles.)

Fabricius ab Aquapendente
De ven. ost.
Fabricius ab Aquapendente, Hieronymus [Girolamo Fabrizio d'Acquapendente]. *De vena-
rum ostiolis.* Padua: Laurentius Pasquatus, 1603. (Also in *Opera.*)
Opera
Fabricius ab Aquapendente, Hieronymus [Girolamo Fabrizio d'Acquapendente]. *Opera
physica.* Padua: R. Miglietti, 1625.

Fernel
Physiol.
Fernel, Jean. *Opera medicinalia; nempe, Phisiologia, Pathologia, & Terapeutica, seu Medendi ratio:
quibus adjecimus De abditis rerum causis: nunc denuo recognita & impressa.* . . . Venetiis, Apud
Franciscum de Portonaris, 1566.

Fontenelle
Œuvres
Fontenelle, Bernard le Bovyer de. *Œuvres diverses.* Amsterdam: s.n., 1742.

Galen
De usu part.
Galen, Claudius. *De usu partium.* In: *Opera* 3, 4:1–366.
Opera
Galen, Claudius. *Opera omnia.* Ed. Carolus Gottlob Kühn. Leipzig: Car. Cnoblochius,
1822.

Gassendi
Disquisitio
Gassendi, Pierre. *Disquisitio Metaphysica seu Dubitationes & Instantiæ adversis Renati Cartesii
Metaphysicam et Responsa.* In: *Opera* 3:269–410. (Orig. publ. Amsterdam, 1644.)
Opera
Gassendi, Pierre. *Opera omnia.* Lyon: Laurentius Anisson, 1658. (Facs. repr. Stuttgart-Bad
Cannstatt: Friedrich Frommann, 1964.)
Syntagma
Gassendi, Pierre. *Syntagma philosophici.* Pt. 1: *Logica.* Pt. 2: *Physica.* Pt. 3: *Ethica.* In: *Opera*
1–2. (Written 1649–1655. On the composition of the *Syntagma* and earlier manuscript
works related to it, see Bloch 1971:xix–xxx.)

Goclenius
Lexicon
Goclenius, Rudolphus. *Lexicon philosophicum* . . . Frankfurt: Matthias Becker, 1613. (Facs.
repr. Hildesheim: Olms, 1964.)

Hippocrates
De nat. pueri
Hippocrates. *Peri gones. De genitura. Peri phusios paidiou. De natura pueri.* Interprete Jo. Gorraeo . . . Accesserunt ejusdem interpretis annotationes in eosdem libellos: in quarum quinquagesimanona, tota temporum pariendi ratio apertissimè explicatur. Parisiis: Ex officina Michaëlis Vascosani, 1545.

Kepler
Ad Vitell.
Kepler, Johannes. *Ad Vitellionem paralipomena,: quibus Astronomiæ pars optica traditur . . . Tractatum luculentum de modo visionis, & humorum oculi vsu, contra opticos & anatomicos. . . .* Francofurti: Apud Claudium Marnium & Hæredes Ioannis Aubrii, 1604.

Lamarck
Philos. zool.
Lamarck, Jean-Baptist-Pierre-Antoine de Monet, chevalier de. *Philosophie zoologique ou exposition des considérations relatives à l'histoire naturelle des animaux. . . .* Nouv. éd. Intro. by Charles Martins. Paris: F. Savy, 1873.

Leonardo da Vinci
Anat. stud.
Leonardo da Vinci. *Corpus of the anatomical studies in the collection of Her Majesty the Queen at Windsor Castle.* Ed. Kenneth D. Keele and Carlo Pedretti. London: Johnson Reprint Co.; New York: Harcourt, Brace, Jovanovich, 1978–1980.

Leurechon
Récréation
Leurechon, Jean. *Récréation mathematique, composée de plusieurs problèmes plaisant et facetieux, en faict d'Arithmeticque, Geometrie, Mechanicque, Opticque, et autres parties de ces belles sciences.* (According to Trevor H. Hall's *Mathematicall Recreations* (1969) Leurechon is not the author. The first part is attributed to Hendrik van Etten and the second part to Denis Henrion and Claude Mydorge). Pont-à-Mousson, Jean Appier Hanzelet, 1624.

Malebranche
Œuvres
Malebranche, Nicolas. *Œuvres.* Ed. Geneviève Rodis-Lewis and Germain Malbreil. Paris: Gallimard, 1979–1992.
RV
Malebranche, Nicolas. *De la recherche de la vérité.* In: *Œuvres* 1 : 3–1126.

Mersenne
Harm. univ.
Mersenne, Marin. *Harmonie universelle, contenant la theorie et la pratique de la musique, où il est traité de la Nature des sons, & des Mouuemens, des Consonances, des Dissonances, des Genres, des Modes, de la Composition, de la voix, des Chants, & de toutes sortes d'Instrumens Harmoniques.* Paris: Sebastien Cramoisy, 1636. (Fasc. repr. of Mersenne's copy, with his marginal notes. Intro. by François Lesure. Paris: Editions du Centre National de la Recherche Scientifique, 1965.)

BIBLIOGRAPHY

Niceron
Perspectives
Niceron, Jean-François. *La perspective curieuse ou Magie artificelle des effets merveilleux de l'optique.* Paris: Chez Pierre Billaine, 1638.

Peiresc
Corr. Gassendi
Peiresc, Nicolas Claude Fabri de. *Lettres à divers: supplément au tome VII de l'édition Tamizey de Larroque et errata.* Publié par Raymond Lebègue, avec la collaboration d'Agnès Bresson. Paris: Editions du Centre National de la Recherche Scientifique, 1985.

Perrault
Méch. des anim.
Perrault, Claude. *De la méchanique des animaux.* In: *Œuv. div.*
Œuv. div.
Perrault, Claude. *Œuvres diverses de physique et de mécanique.* Leiden: P. van der Aa, 1721.

Piles
Conversations
Piles, Roger de. *Conversations sur la connaissance de la peinture et sur le jugement qu'on doit faire des tableaux.* Genève, Slatkine Reprints, 1970. (Orig. publ. Paris: Nicolas Langlois, 1677.)

Poisson
Commentaire
Poisson, Nicolas. *Commentaire ou remarques sur la méthode de R. Descartes.* Vendôme: S. Hip, 1670. (Repr. New York: Garland, 1987.)

Quesnay
Observations
Quesnay, François. *Observations sur les effets de la saignée tant tous les maladies . . . fondées sur les loix de l'hidrostatique; avec des remarques critiques sur le Traite de l'usage des differentes sortes de saignées de monsieur Silva.* Paris: C. Osmont, 1730.
Traité
Quesnay, François. *Traité des effets et de l'usage de la saignée.* Paris: d'Houry père, 1750. (Revised version of *Observations* and *L'Art de guérir par la saignée,* 1736.)

Regius
Fund. phys.
Regius, Henricus. *Fundamenta physices.* Amsterdam: Ludovicus Elzevirius, 1646.

Schultens
In mem. Boerhavii
Schultens, Albert. *Oratio academica. In memoriam Hermanni Boerhaavii. . . . Ex decreto rectoris magnifici et senatus academici habita die IV Novembris, an MDCCXXXVIII.* Leiden: J. Luzac, 1738.

Stahl
Diss. inaug.
Stahl, George Ernst. *Dissertatio inauguralis medica de medicina medicinæ curiosæ.* Halae Magdeburgicæ: Literis Christiani Henckelii, 1714.

BIBLIOGRAPHY

Œuvres

Stahl, George Ernst. *Œuvres médico-philosophiques et pratiques de G. E. Stahl.* . . . Traduites et commentées par T. Blondin . . . augmentées d'arguments et de réflexions philosophiques et médicales, par L. Boyer . . . contenant enfin de remarquables travaux inédits de M. Tissot. Paris: J.-B. Baillière et fils, 1859.

Suárez

De angelis

Suárez, Franciscus. *De angelis.* In: *Opera,* v.2. (Orig. publ. 1620.)

De An.

Suárez, Franciscus. *De anima.* In: *Opera,* v. 3. (Orig. publ under the title *Partis secundæ Summæ theologiæ Tomus alter, complectens tractatum secundum de Opere sex dierum ac tertium de Anima.* Lugdunum [Lyon]: Jacques Cardon & Pierre Cavellat, 1621.)

Disp.

Suárez, Franciscus. *Disputationes metaphysicæ.* Hildesheim: Olms, 1965. (Reprint of v. 25 – 26 of the *Opera.*)

Opera

Suárez, Franciscus. *Opera omnia.* Ed. D. M. André. Paris: L. Vivès, 1856.

Sylvius

Isagoge

Sylvius, Iacobus (Jacques Dubois). *In Hippocratis et Galeni physiologiae partem anatomicam Isagoge. Denuo per Alexandrum Arnaudum diligentissimè castigata.* Parisiis: Aegidium Gorbinum, 1560. (*Manuscripta: microfilms of rare and out-of-print books:* list 15, no.33)

Opera

Sylvius, Iacobus (Jacques Dubois). *Opera medica, iam demum in sex partes digesta, castigata et indicibus necessariis instructa.* Adiuncta est eiusdem vita et icon, opera et studio Renati Moraei. Genevae: Iacobi Chouet, 1630. (*Manuscripta: microfilms of rare and out-of-print books:* list 15, no.34)

Thomas Aquinas

In de An.

Pirotta

Thomas Aquinas. *In Aristotelis librum de anima commentarium.* Ed. A. M. Pirotta. Turin: Marietti, ²1936.

Opera

Leonina

Thomas Aquinas. *Opera omnia.* Rome: Commissio Leonina, 1882–.

Parma

Thomas Aquinas. *Opera omnia.* Parma: Petrus Fiaccodorus, 1852–1873. Facs. repr. New York: Musurgia, 1949.

ST

Thomas Aquinas. *Summa theologiæ.* In: *Opera* Parma v. 1–3; Leonina v. 4–12.

Toletus

In de An.

Toletus, Franciscus. *Commentaria una cum quæstionibus in III libros De anima.* In: *Opera,* v. 3. (Orig. publ. Köln, 1575, subsequent separate editions to 1625.)

In de Gen.
Toletus, Franscicus. *Commentaria una cum quæstionibus in II libros De generatione et corruptione.* In: *Opera,* v. 5. (Orig. publ. 1575, subsequent separate editions to 1602.)

In Log.
Toletus, Franciscus. *Commentaria una cum Quæstionibus in universam Aristotelis Logicam.* In: *Opera,* v. 2. (Orig. publ. 1572, subsequent editions to 1616.)

In Phys.
Toletus, Franciscus. *Commentaria unà cum Quæstionibus in octo libros Aristotelis de Physica auscultatione.* In: *Opera,* v. 4. (Orig. publ. 1572, with subsequent separate editions to 1589.)

Opera
Toletus, Franciscus. *Opera omnia philosophia.* Intro. by Wilhelm Risse. Hildesheim: Georg Olms, 1985. (Facs. repr. of Köln 1615–1616 ed. in 5v.)

Vesalius
De fabrica
Vesalius, Andreas. *De humani corporis fabrica libri septem.* Basle: Oporinus, 1543, [2]1555.

Zonca
Novo Teatro
Zonca, Vittorio. *Novo teatro di machine et edificii per varie et sicure operationi [. . .].* Padua: Francesco Bertelli, 1607, [4]1656. (Facs. repr (of 1607 ed.) = *Libri rari: collezione di ristampe con nuovi apparati,* 7. Milano: Il Polifilo, 1985.)

SECONDARY SOURCES

Albistur & Armogathe 1977
Albistur, Maïté and Daniel Armogathe. *Histoire du féminisme français du Moyen Âge à nos jours.* Vol. 1. Paris: Éditions des Femmes, 1977.

Baigrie 1996a
Baigrie, Brian S. "Descartes's scientific illustrations and 'la grande mecanique de la nature'." In: Baigrie 1996b:86–134.

Baigrie 1996b
Baigrie, Brian S., ed. *Picturing knowledge: Historical and philosophical problems concerning the use of art in science.* Toronto: University. of Toronto Press, 1996.

Baker & Morris 1996
Baker, Gordon, and Katherine J. Morris. *Descartes' dualism.* London/New York: Routledge, 1996.

Baltrusaitis 1984
Baltrusaitis, Jurgis. *Anamorphoses ou Thaumaturgus opticus.* Paris: Flammarion, 1984. (The second volume of his *Perspectives dépravées.* Orig. publ. under the title *Anamorphoses ou perspectives curieuses* (Paris: Olivier Perrin, 1955).)

Baxandall 1985
Baxandall, Michael. *Patterns of intention: On the historical explanation of pictures.* New Haven/London: Yale University Press, 1985.

BIBLIOGRAPHY

Beaune 1980
Beaune, Jean-Claude. *L'automate et ses mobiles*. Paris: Seuil, 1980.

Becq 1982
Becq, Annie. "La métaphore de la machine dans le discours esthétique de l'âge classique." *Revue des sciences humaines* 58(Apr–Oct 1982):269–278.

Belgioioso 1990
Belgioioso, G., G. Cimino et al., eds. *Descartes: il metodo e i saggi*. Rome: Istituto della Enciclopedia Italiana, 1990.

Beyssade & Marion 1994
Beyssade, Jean-Marie, and Jean-Luc Marion. *Descartes. Objecter et répondre*. Avec la collaboration de Lia Levy. Paris: Presses Universitaires de France, 1994.

Biard & Rashed 1997
Biard, Joël, and Roshdi Rashed. *Descartes et le Moyen Âge*. Paris: J. Vrin, 1997. (Études de philosophie médiévale)

Bitbol-Hespériès 1988
Bitbol-Hesperies, Annie. "Le principe de vie dans les 'passions de l'âme'." *Revue philosophique* 178(1988):415–431.

Bitbol-Hespériès 1990
Bitbol-Hespériès, Annie. *Le Principe de vie chez Descartes*. Paris: J. Vrin, 1990.

Bitbol-Hespériès 1994
Bitbol-Hespériès, Annie. "Réponse à Vere Chappell. L'union substantielle." In: Beyssade & Marion 1994:427–448.

Bitbol-Hespériès 1996
Bitbol-Hespériès, Annie. "Introduction." In: *Descartes. Le Monde/L'Homme* iii–liii.

Bitbol-Hespériès 1998
Bitbol-Hespériès, Annie. "La médecine et l'union dans la Méditation sixième." In: Kolesnik-Antoine 1998a:18–36.

Bloch 1971
Bloch, Olivier René. *La philosophie de Gassendi. Nominalisme, matérialisme, et métaphysique*. La Haye: Martinus Nijhoff, 1971. (*Archives internationales d'histoire des idées*, 38.)

Bordo 1999
Bordo, Susan, ed. *Feminist interpretations of René Descartes*. University Park, Pennsyl.: Pennsylvania State University Press, 1999. (*Re-reading the canon*)

Burnyeat 1992
Burnyeat, Miles. "Is an Aristotelian philosophy of mind still credible? (A draft)." In: Nussbaum & Rorty 1992:15–26.

Canguilhem 1975
Canguilhem, Georges. *La connaissance de la vie*. Paris: J. Vrin, ²1975. (Orig. publ. 1952.)

Canguilhem 1977
Canguilhem, Georges. *La formation de la concept de réflexe aux xviie et xviiie siècles*. Paris: J. Vrin, [2]1977.

Carter 1983
Carter, Richard B. 1983. *Descartes' medical philosophy. The organic solution to the mind-body problem*. Baltimore: Johns Hopkins University Press.

Carter 1985
Carter, Richard B. 1985. "Descartes's bio-physics." *Philosophia naturalis* 22(1985):223–249. (Repr. in Moyal 1991.)

Cavaillé 1991
Cavaillé, Jean-Pierre. Descartes. *La fable du monde*. Paris: École des Hautes Études en Sciences Sociales/J. Vrin, 1991.

Chappell 1992
Chappell, Vere, ed. *René Descartes*. New York/London: Garland, 1992.

Chappell 1994
Chappell, Vere. "L'homme cartésien." In: Beyssade & Marion 1994:403–426.

Clarke 1999
Clarke, Stanley. "Descartes' 'Gender'." In: Bordo 1999:82–102.

Cooper 1987
Cooper, John M. "Hypothetical necessity and natural teleology." In Gotthelf & Lennox 1987:243–274.

Cottingham 1994
Cottingham, John, ed. *Reason, will, and sensation: Studies in Cartesian metaphysics*. Oxford: Clarendon, 1994.

Cult. delle Macchine 1989
La cultura delle macchine. Dal medieoevo all rivoluzione industriale nei documenti dell'archivio storico AMMA. Turin: Umberto Allemandi, 1989.

Cummins 1983
Cummins, Robert. *The nature of psychological explanation*. Cambridge, Mass.: MIT/Bradford, 1983.

Daremberg 1870
Daremberg, Charles. *Histoire des sciences médicales, comprenant l'anatomie, la physiologie, la médicine, la chirurgie et les doctrines de pathologie générale*. Paris: J.-B. Baillière et fils, 1870.

Daston & Park 1998
Daston, Lorraine and Katherine Park. *Wonders and the order of nature, 1150–1750*. New York: Zone, 1998.

Des Chene 1996
Des Chene, Dennis. *Physiologia. Natural philosophy in late Aristotelian and Cartesian thought*. Ithaca, N.Y.: Cornell University Press, 1996.

BIBLIOGRAPHY

Des Chene 1997
Des Chene, Dennis. "L'immatérialité de l'âme: Suárez et Descartes." In: Biard & Rashed 1997:319–328.

Des Chene 2000
Des Chene, Dennis. *Life's Form: Late Aristotelian Conceptions of the Soul.* Ithaca, N.Y.: Cornell University Press, 2000.

Detienne 1967/1994
Detienne, Marcel. *Les Maîtres de vérité dans la Grèce archaïque.* Paris: La Découverte, ²1990; Pocket, 1994. (Orig. publ. 1967.)

Deutsch 1951
Deutsch, Karl. "Mechanism, organism, and society: Some models in natural and social science." *Philosophy of science* 18 (1951):230-252.

Doyon & Liaigre 1957
Doyon, A., and L. Liaigre. "Méthodologie comparée du biomécanisme et de la mécanique comparée." *Dialectica* 10.4(1956):292–335.

Dretske 1981
Dretske, Fred. *Knowledge and the flow of information.* Cambridge, Mass.: MIT/Bradford, 1981.

Duchesneau 1982
Duchesneau, François. *La physiologie des Lumières. Empirisme, modèles et théories.* The Hague [etc.]: Martinus Nijhoff, 1982. (*Archives internationales d'histoire des idées, 95*)

Espinas 1903
Espinas, Alfred. "L'organisme ou la machine vivante en Grèce, au IVe siècle avant J. C." *Revue de métaphysique et de morale* 11(1903):703–715.

Fritscher & Brey 1994
Fritscher, Bernhard, and Gerhard Brey, eds. *Cosmographica et geographica: Festschrift für Heribert M. Nobis zum 70. Geburtstag.* München: Institut für Geschichte der Naturwissenschaften, 1994. (*Algorismus, Heft, 13*)

Fuchs 1992
Fuchs, Thomas. *Die Mechanisierung des Herzens: Harvey und Descartes, der vitale und der mechanische Aspekt des Kreislaufs.* Frankfurt am Main: Suhrkamp, 1992.

Gabbey 1990
Gabbey, Alan. "Explanatory structures and models in Descartes' physics." In Belgioioso 1990, 1:273–286.

Galluzzi 1987
Galluzzi, Paolo. "Leonardo da Vinci: From the 'elementi macchinali' to the man-machine." *History and Technology* 4 (1987):235–265.

Garber 1993
Garber, Daniel. "Descartes and experiment in the Discourse and Essays." In: Voss 1993: 288–310.

Gaukroger 1995
Gaukroger, Stephen. Descartes. *An intellectual biography.* Oxford: Clarendon, 1995.

Giglioni 1995
Giglioni, Guido. "Automata compared: Boyle, Leibniz and the debate on the notion of life and mind." *British Journal for the History of Philosophy* 3.2 (1995): 249–278.

Gilson 1984
Gilson, Étienne. *Études sur le rôle de la pensée médiévale dans la formation du système cartésien.* Paris: Vrin, ⁵1984. (Orig. publ. 1930.)

Gontier 1991
Gontier, Thierry. "Les animaux-machines chez Descartes: modèle ou réalité?" *Corpus: Revue de Philosophie* 16/17(1991): 3–16.

Gotthelf & Lennox 1987
Gotthelf, Allan, & James G. Lennox, eds. *Philosophical issues in Aristotle's biology.* Cambridge: Cambridge University Press, 1987.

Gracia 1994a
Gracia, Jorge. "Francis Suárez." In: Gracia 1994b: 475–510.

Gracia 1994b
Gracia, Jorge J. E., ed. *Individuation in Scholasticism. The later Middle Ages and the Counter-Reformation, 1150–1650.* Albany, N.Y.: State University of New York Press, 1994.

Grene 1993
Grene, Marjorie. "The heart and blood: Descartes, Plemp, and Harvey." In: Voss 1993: 324–336.

Grimaldi 1978
Grimaldi, Nicolas. *L'expérience de la pensée dans la philosophie de Descartes.* Paris: J. Vrin, 1978.

Grosholz 1991
Grosholz, Emily. *Cartesian method and the problem of reduction.* Oxford: Clarendon Press, 1991.

Gueroult 1968
Gueroult, Martial. *Descartes selon l'ordre des raisons.* Paris: Aubier, 1968. 2v. (Eng. trans. by Roger Ariew. *Descartes' philosophy interpreted according to the order of reasons.* Minneapolis: University of Minnesota Press, 1984–1985.)

Gueroult 1970
Gueroult, Martial. *Études sur Descartes, Spinoza, Malebranche, et Leibniz.* Hildesheim, New York, G. Olms, 1970. (*Studien und Materialien zur Geschichte der Philosophie,* Bd. 5)

Guyénot 1941
Guyénot, Émile. *Les sciences de la vie aux XVIIᵉ et XVIIIᵉ siècles. L'idée d'évolution.* Paris: Albin-Michel, 1941.

Hall 1969

Hall, Thomas S. *Ideas of life and matter. Studies in the history of general physiology, 600 B.C.– 1900 A.D.* v. 1: *From Pre-Socratic times to the Enlightenment.* Chicago: University of Chicago Press, 1969.

Hankinson

Hankinson, R J. "Galen and the best of all possible worlds." *Classical Quarterly* n.s.39 (1989):206–27.

Hoffman 1987

Hoffman, Paul. "The unity of Descartes's man." *Philosophical review* 95 (1986):339–370. (Repr. in Moyal 1991, 3:168–192, and in Chappell 1992, 2:19–50.)

P. Hoffman 1982

Hoffman, P. "De quelques aspects de la représentation du vivant chez Descartes, Borelli et Stahl." *Revue des sciences humaines* 58 (1982):199–211.

Huffman 1988

Huffman, William H. *Robert Fludd and the end of the Renaissance.* London: Routledge, 1988.

Kambouchner 1995

Kambouchner, Denis. *L'Homme des passions. Commentaires sur Descartes.* I: *Analytique.* II: *Canonique.* Paris: Albin Michel, 1995.

Kolesnik-Antoine 1998A

Kolesnik-Antoine, Delphine. "'Comme un pilote en son navire': Arnaud lecteur de la VIème Méditation." In: Kolesnik-Antoine 1998b: 100–128.

Kolesnik-Antoine 1998B

Kolesnik-Antoine, Delphine, ed. *Union et distinction de l'âme et du corps: lectures de la viᵉ Méditation.* Préface de Pierre-François Moreau. Paris: Éditions Kimé, 1998.

Kraye & Stone 2000

Kraye, Jill, and M. W. F. Stone, eds. *Humanism and Early Modern Philosophy.* London: Routledge.

Lindberg 1970

Lindberg, David C. *John Pecham and the science of optics. Perspectiva communis.* Madison: University of Wisconsin Press, 1970.

Lloyd 1992

Lloyd, G. E. R. "Aspects of the relationship between Aristotle's psychology and his zoology." In Nussbaum & Rorty 1992:147–167.

Maritain 1924

Maritain, Jacques. *Trois réformateurs: Luther, Descartes, Rousseau.* Nouv. éd. Paris: Plon, 1937. ("Descartes ou l'incarnation de l'ange" is pp. 73–128 and 293–306. Orig. publ. in *Revue universelle* 15 Déc. 1924.)

Mendelsohn 1964

Mendelsohn, Everett. *Heat and life: The development of the theory of animal heat.* Cambridge: Harvard University Press, 1964.

BIBLIOGRAPHY

Menn forthcoming
Menn, Stephen. "On Dennis Des Chene's *Physiologia.*" *Perspectives on Science.*

Mesnard 1937
Mesnard, Pierre. "L'esprit de la physiologie cartésienne." *Archives de philosophie* 13 (1937):181–220.

Michael & Michael 1989
Michael, Emily & Fred S. Michael. "Corporeal ideas in seventeenth-century psychology." *Journal of the history of ideas* 50 (1989):31–48.

Moyal 1991
Moyal, Georges J. D., ed. *René Descartes. Critical assessments.* London/New York: Routledge, 1991.

Nussbaum & Rorty 1992
Nussbaum, Martha, and Amélie Oksenberg Rorty, eds. *Essays on Aristotle's De Anima.* Oxford: Clarendon, 1992.

Osler 1996
Osler, Margaret. "From Immanent Natures to Nature as Artifice: The Reinterpretation of Final Causes in Seventeenth-Century Natural Philosophy." *The Monist* 79 (1996): 388–407.

Osler 2000
Osler, Margaret. "Renaissance Humanism, Lingering Aristotelianism and the New Natural Philosophy: Gassendi on Final Causes". In Kraye & Stone 2000:193-208.

Osler forthcoming
Osler, Margaret. "Whose Ends? Teleology in Early Modern Natural Philosophy." *Osiris* 16 (2001).

Perler 1996
Perler, Dominik. *Repräsentation bei Descartes.* Frankfurt am Main: V. Klostermann, 1996. (*Philosophische Abhandlungen,* Bd. 68)

Pichot 1993
Pichot, André. *Histoire de la notion de vie.* Paris: Gallimard, 1993.

Quine 1960
Quine, W. V. O. *Word and object.* Cambridge, Mass.: MIT, 1960.

Radner 1971
Radner, Daisie. "Descartes' notion of the union of mind and body." *Journal of the history of philosophy* 9 (1971):159–170. (Repr. in Moyal 1991, 3:276–288.)

Reuleaux 1874–5/1876
Reuleaux, F. *The kinematics of machinery. Outlines of a theory of machines.* Trans. Alex. B. W. Kennedy. London: Macmillan and Co., 1876. (Translation of *Theoretische Kinematik,* 1874–1875.)

Ritchie 1964
Ritchie, A. M. "Can animals see? A Cartesian query." *Proceedings of the Aristotelian Society* 64 (1964):221–242. (Repr. in Moyal 1991, 4:280–296.)

Rodis-Lewis 1950
Rodis-Lewis, Geneviève. *L'Individualité selon Descartes*. Paris: J. Vrin, 1950. (*Bibliothèque de l'histoire de la philosophie*)

Rodis-Lewis 1959
Rodis-Lewis, Geneviève. *Lettres à Regius et Remarques sur l'Explication de l'Esprit humain*. Paris: J. Vrin. 1959.

Rodis-Lewis 1971
Rodis-Lewis, Geneviève. *L'Œuvre de Descartes*. Paris: J. Vrin, 1971.

Rodis-Lewis 1990
Rodis-Lewis, Geneviève. *L'anthropologie cartésienne*. Paris: PUF, 1990.

Roger 1971
Roger, Jacques. *Les sciences de la vie dans la pensée française du xviiie siècle. La génération des animaux de Descartes à l'*Encyclopédie. Seconde éd. complétée. Paris: Armand Colin, 1971. (Orig. publ. 1963.)

Romanowski 1974
Romanowski, Sylvie. *L'illusion chez Descartes. La structure du discours cartésien*. Paris: Klinck-sieck, 1974. (*Critères, 3*)

Rothschuh 1968
Rothschuh, Karl. Physiologie. *Der Wandel ihrer Konzepte, Probleme und Methoden vom 16. bis 19. Jahrhundert*. Freiburg/München: Karl Alber, 1968.

Sawday 1995
Sawday, Jonathan. *The body emblazoned. Dissection and the human body in Renaissance culture*. London/New York: Routledge, 1995.

Schlanger 1971
Schlanger, Judith. *Les métaphores de l'organisme*. Paris: J. Vrin, 1971.

Segre 1994
Segre, Michael. "Ramellis 'machine': Realitat oder Fiktion?." In: Fritscher & Brey 1994, 1:367–372.

Siraisi 1990
Siraisi, Nancy. *Medieval and early Renaissance medicine: An introduction to knowledge and practice*. Chicago: University of Chicago Press, 1990.

Verbeek 1988
Verbeek, Theo. 1988. "Introduction." In Descartes & Schoock *Querelle*, 19–66.

Verbeek 1992
Verbeek, Theo. *Descartes and the Dutch. Early reactions to Cartesian philosophy, 1637–1650*. Carbondale: Southern Illinois University Press, 1992.

Verbeek 1993

Verbeek, Theo. *Descartes et Regius. Autour de l'Explication de l'esprit humain.* Amsterdam/
Atlanta, GA: Rodopi, 1993. (*Colloques du Centre franco-néerlandais de recherches cartésien-
nes,* 1)

Voss 1993

Voss, Stephen, ed. *Essays on the philosophy and science of René Descartes.* Oxford: Oxford University Press, 1993.

Voss 1994

Voss, Stephen. "Descartes: The end of anthropology." In: Cottingham 1994:

Wear 1983

Wear, Andrew. "Harvey and the way of the anatomists." *History of science* 21 (1983):223–
249.

Whiting 1992

Whiting, Jennifer. "Living bodies." In: Nussbaum & Rorty 1992:75–92.

Wilson 1997

Wilson, Mark. "Mechanism and Fracture in Cartesian Physics." *Topoi* 16.2 (1997):141–
152.

Yablo 1997

Yablo, Stephen. "Wide causation." *Philosophical perspectives* (*Mind, causation, and world,* ed.
James E Tomberlin) 11 (1997):251–281. (Published as a supplement to *Noûs.*)

Index